23 Advances in Polymer Science

Fortschritte der Hochpolymeren-Forschung

With 28 Figures

Springer-Verlag
Berlin Heidelberg GmbH 1977

Editors

ISBN 978-3-662-15474-8 ISBN 978-3-540-38030-6 (eBook)
DOI 10.1007/978-3-540-38030-6

Library of Congress Catalog Card Number 61-642

Originally published by Springer-Verlag Berlin Heidelberg New York in 1977
Softcover reprint of the hardcover 1st edition 1977

43210

Contents

Polymeric Reagents Polymer Design, Scope, and Limitations

Walter Heitz
Fachbereich Physikalische Chemie, Polymere, Philipps-Universität, D-3550 Marburg, Germany

Table of Contents

1. Introduction

The use of polymeric reagents dates back to the mid-30, s when ion exchangers made by condensation processes became available (*1*). Their structure was greatly improved by introducing ion exchangers based on styrene/divinylbenzene in 1944 (*2*). But it was not until the pioneering work of Merrifield (*3*) in 1963 on peptide synthesis on polymer supports when the advantage of polymer-supported reactions were fully recognized. This was despite the work on electron-exchange resins beginning in 1949 [Cassidy (*4*), Manecke (*5*)] and enzymes fixed on polymers dating back to 1951 (*6*) [Campbell *et al.* 1951 (*7*), Grubhofer and Schleith 1954 (*8*)].

In this paper the strongly increasing possibilities for using polymeric supports (*9, 10*) for organic synthesis is not discussed in detail, but rather the polymer backbone is brought into focus and our knowledge which is important for the construction and use of polymeric supports is summarized.

2. The Macrostructure

2.1. Possibilities

The advantage of polymer supports for organic synthesis may have different causes:

1) Ease of separation. In most applications filtration of the reagent from the reaction mixture is a decisive gain.
2) Reagents attached to polymers are nontoxic and odorless. Dangerous materials will be easy to handle.
3) Effect of environment. Functional groups attached to polymers are in another local environment than in free solution. The kinetics of a reaction may be influenced by this effect.
4) Effect of polymer backbone. We can make use of local concentration effects.

In principal, any linear polymer with functional groups may be regarded as a polymeric reagent. With increasing molecular weight they will be easier to separate but at $M \approx 10^6$ making concentrated solutions (10–20%) is time-consuming and their viscosity presents problems. When crosslinking sites are introduced, the molecular coil size and, as a consequence, the viscosity is decreased. So practically speaking, "infinite molecular weights" can be handled in the form of microgels and "crosslinked polymers". In all these cases the functional groups are threated on the individually solvated polymer molecule. Quite another approach is to make the crosslinking density so high that an unpenetrable particle results. Reactions can take place only at the surface which correspondingly has to be sufficiently high in density. Again,microgels and "macroporous polymers" are possible solutions.

2.2. Linear Polymers

All the advantages of polymer supports are in principal fulfilled even if we use linear polymers ($M \simeq 10^4-10^5$). This enables working in a homogeneous phase but there is no possibility of removing the polymer just by filtration. Precipitation and more likely ultrafiltration are to be applied. As ultrafiltration is becoming increasingly convenient, the use of linear polymers is of growing interest (*11, 12*). Compared with crosslinked polymers linear polymers have the advantage that diffusion restrictions are small, high conversions can be obtained (*13*), and all functional groups are of similar accessibility. Problems arising from the pore-size distribution and, as a consequence, the limited yield for substrates of larger molecular size are restricting factors for the use of crosslinked polymers. With linear polymers a polymeranalogous reaction (*i.e.* reaction not effecting the size of the polymer backbone) usually proceeds to a higher extent.

For example, if a brominated polystyrene is treated with K_2Se and hydrolyzed we obtain – starting from linear polystyrene – a polymer with about 30% Se, whilst the crosslinked polymer – even with the low amount of 0.5% DVB – gives about 20% Se under identical conditions (*14*). Separation problems are not always combined with linear polymers. So N-Chloro-nylons are readily soluble in many solvents and can be used for oxidation and chlorination (*15*). Unsubstituted nylon regenerated during this reaction is insoluble in most solvents; therefore it precipitates during the reaction.

2.3. Crosslinked Polymers

The predominant method used to make a polymeric reagent is to synthetize a crosslinked polymer. This network structure can be greatly varied and may be of two extremes:

1) A random distribution of crosslinking sites throughout the polymer particle.
2) An accumulation of crosslinking sites resulting in unpenetrable cores bound together by the polymer. A polymer which may have a porous structure even in the dry state.

Characterization of the crosslinked polymer in the dry state [apparent density (*16*), surface measurement by N_2-adsorption (*17, 18*), Hg-intrusion for measurement of the pore volume (*18*)] is not conclusive for the properties as polymeric reagent,

However, extensive knowledge about the porous structure and the accessibility of different regions in the polymer network can be obtained by gel-permeation chromatography (GPC) (*20*). GPC is used in an inverse mode. Well-characterized samples are keys for the pore structure.

In GPC, a molecule too large to enter any pore will be eluted with the interstitial volume. Smaller molecules will be increasingly delayed the more pore volume is accessible to them (Fig. 1).

Ignoring specific interaction effects and the influence of solvent on molecular-coil size, the usual plot of log M vers. ν_E is equivalent to the integral pore-size distribution in an arbitrary scale [M = pore size; $\nu_E = I$ (pore size)][1]. For a given substrate with a molecular weight M_1 the available part of the pore volume is a/ν_I. Functional

[1] Actually this is a simplification. As a result of random fluctuations of pore radii (*21, 22*) larger pores in the interior are not available without severe diffusional restrictions.

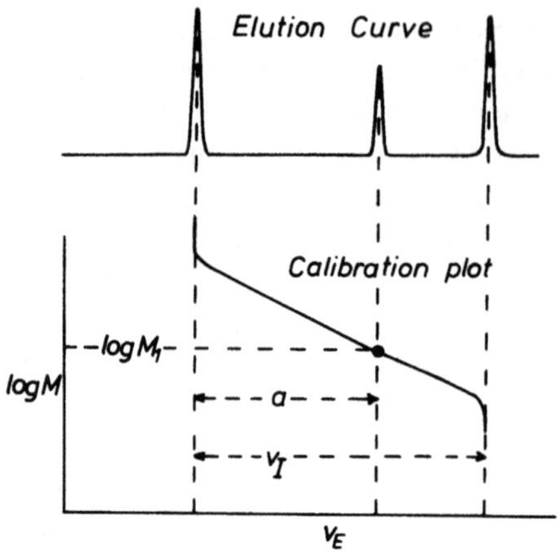

Fig. 1. Schematic representation of GPC

groups in smaller pores react very slowly or are even unable to react with a molecule of this size.

The amount of information that we obtain by GPC is best demonstrated by a separation of oligomers (Fig. 2). Each polymerization degree produces a peak. The difference in elution volume between *e.g.* $M = 2138$ ($n = 20$) and $M = 2242$ ($n = 21$) is caused by a small increment in additionally available pore volume.

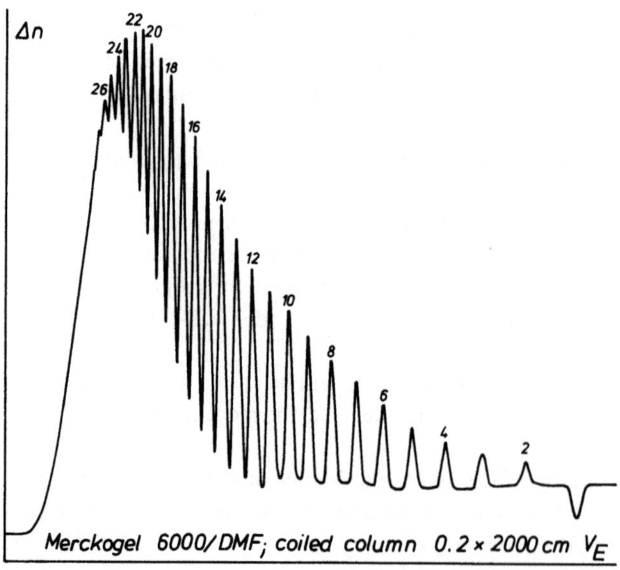

Fig. 2. Separation of oligostyrene 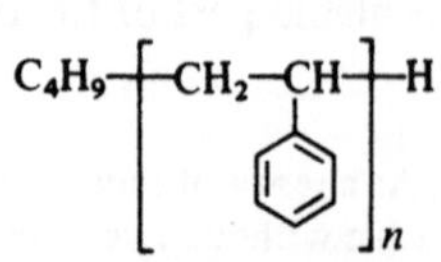 by GPC Figures at the peaks correspond to the polymerization degree n

2.3.1. Randomly Crosslinked Polymers

Crosslinking a linear polymer in absence of a solvent or in presence of a good solvent will result in a uniform distribution of crosslinking density.

The predominant method used to prepare a crosslinked polymer support is to polymerize a monovinylcompound with a small amount of a divinylcompound by radical initiators. This enables the use of suspension techniques and spherical particles are therefore available.

The crosslinking sites will be randomly distributed if we use low amounts of crosslinking agent, work in the absence of a solvent, and take a monomer mixture with appropriate copolymerization behavior. If the reactivities of the two double bonds in a divinyl compound BB are independent of each other and equal, then the copolymerization parameter r_b of an analog monovinyl compound B is a good measure for the crosslinking density obtainable with monomer A ($r_a = k_{aa}/k_{ab}$; $r_b = k_{bb}/k_{ba}$; k_{ab} rate constant of the addition of monomer B to the chain end ⁓A·, etc). If both copolymerization parameters are small, that is, if we have a strong tendency to alternating polymers, then the crosslinking agent, which is usually present in small amounts, is consumed in the first stage and a very inhomogeneous network is formed. If r_A is small and r_B large, we have a similar effect. $r_B \ll r_A$ indicates that BB will be very ineffective as crosslinking agent, which is of special importance if large pores should be formed.

When polymerizing a monovinyl compound with a small amount of a divinyl compound, the amount of crosslinking agent determines the pore size. The pore size will increase with decreasing amounts of crosslinking agent (Fig. 3).

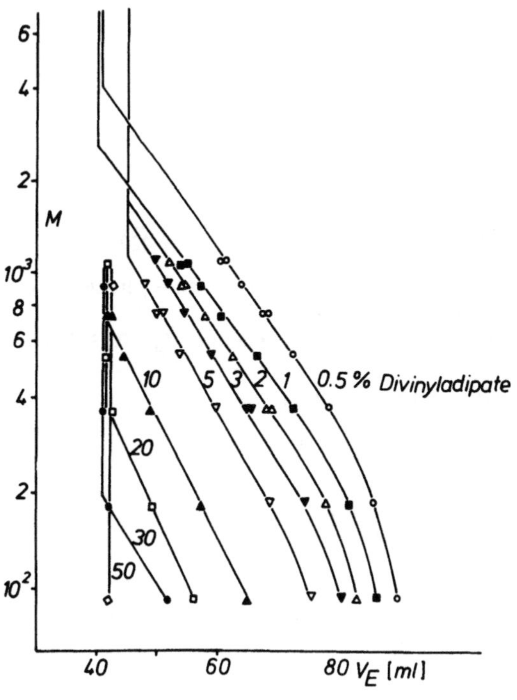

Fig. 3. GPC elution behavior of polyvinyl acetate gels crosslinked with different amounts of divinyl adipate

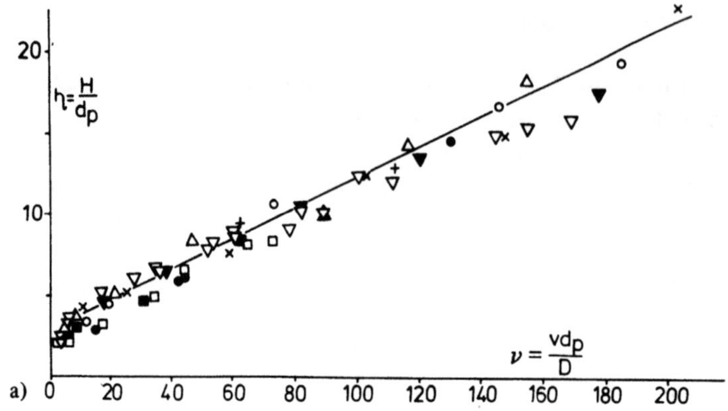

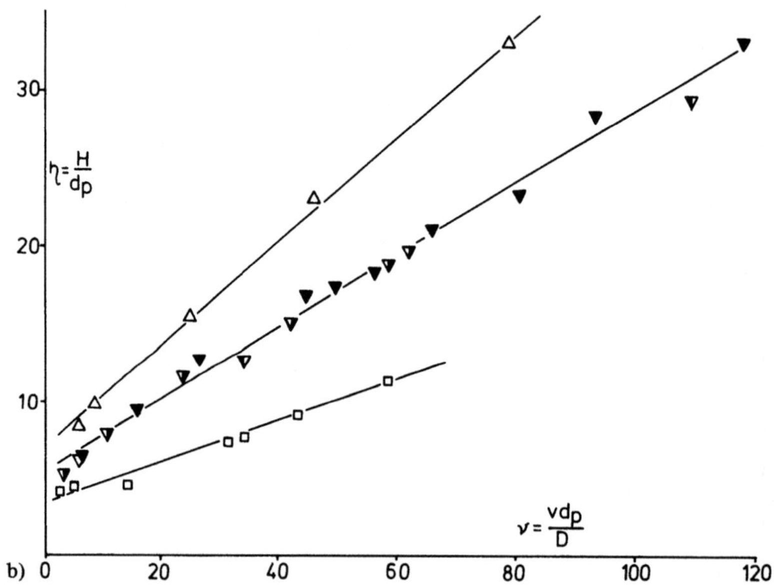

Fig. 4. Reduced separation efficiency in GPC (H – equivalent plate height, d_p – particle size of the gel, v – linear elution rate, D – diffusion coefficient of the sample)
a) Polyvinyl acetate gel (8% butandiol divinyl ether) /THF / oligophenylenes. $\bar{d}_p$ = 0.0382 cm ■, ▲, ▲, +, ●; d_p = 0.0191 cm □, △, △, x, ○
b) Polystyrene gel (10% divinylbenzene) /THF / obligophenylenes
$\bar{d}_p$ = 0.0169 cm □,▼,△ ; $\bar{d}_p$ = 0.0153 cm ▼ ; $\bar{d}_p$ = 0.0416 ▼

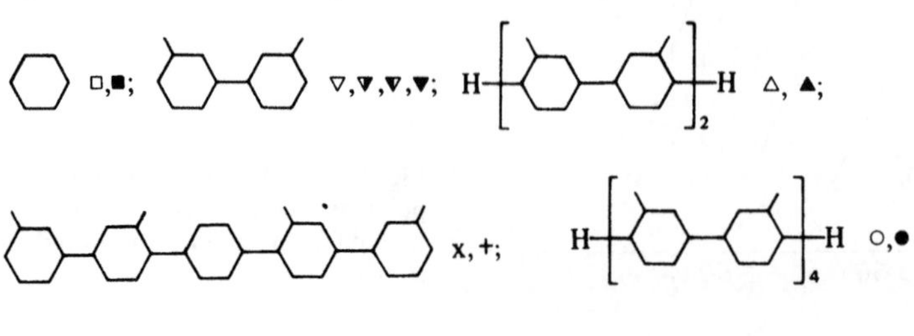

The elution behavior in GPC shows that the working range of these gels indeed shifts to higher molecular weights with decreasing amount of divinyl compound. When taking the gel 3% crosslinked as an example, we see that a substrate with a molecular weight of 500 can penetrate less than half of the total pore volume. There is another problem ruling out the use of high crosslinking densities: the diffusional restriction. Diffusional restrictions manifest themselves in GPC by an additional peak-broadening. Using reduced quantities (*23, 24*) peak-spreading in GPC can be plotted on a master curve (*25*) (Fig. 4a). The values are on a straight line independent of the chemical nature of the gel and within certain limits of the particle size, flow rate, molecular weight of the sample as well as crosslinking density[2]. But an increase of the amount of crosslinking agent to more than approximately 3% will give quite a different picture (Fig. 4b). The flow-rate dependence of peak-spreading becomes greater, the higher the molecular size of the substrate; and finally we get skewed elution curves and a dependence of elution volume on flow rate. The increase in diffusion restrictions with higher crosslinking densities has been shown more recently by Regen (*26*) using electron paramagnetic resonance (EPR) (Fig. 5).

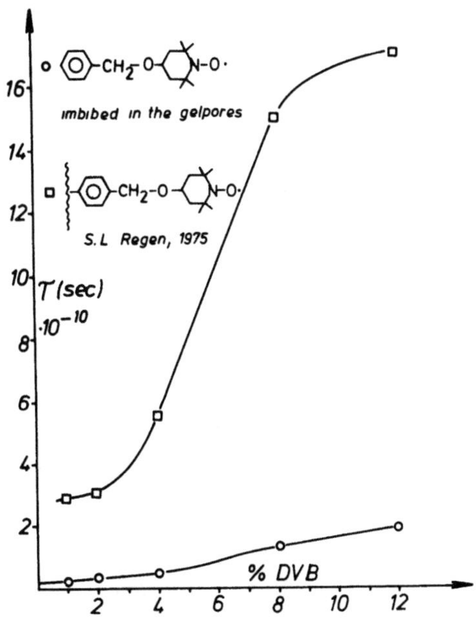

Fig. 5. EPR rotational correlation time of nitroxides in polystyrene crosslinked with different amounts of divinyl benzene (DVB)

The rotational correlation times of a benzene solution of a nitroxide imbibed in polystyrene of varied crosslinking density and the analogous compound covalently attached to the chains of the crosslinked polystyrene were measured. In the latter case the modified polystyrene matrix was also equilibrated with benzene. An estimate of the influence of crosslinking density on the "internal viscosity" can be obtained

---

[2] Butandiol divinylether is very ineffective in crosslinking vinylacetate, so even using 8% the crosslinking density is low.

using the relationship $\tau \simeq \eta/T$, where η is the macroscopic solution viscosity and T is the temperature. The correlation times for the nitroxide in the polymer channels are twice as high for a 2% crosslinked polystyrene as compared to free solution and increase over the whole range by an order of magnitude. The nitroxide fixed at the polymer backbone has a much lower mobility and the correlation times decrease drastically if the amount of DVB is higher than 4%. These results emphasize the fact that the amount of crosslinking agent should be kept small. Lowering the crosslinking density will increase the swelling, resulting in soft gels which are not easy to handle. For PS there is an experimental limit at about 0.5% DVB. Lower crosslinked gels are difficult to filter[3]. Under drastic reaction conditions gels of very low crosslinking density can get soluble by degradation. Ultrafiltration can extend this range but discards the elementary separation step by a more sophisticated one.

The most used polymer support is the polystyrene crosslinked with 2% DVB. In Figure 6 the GPC elution behavior of this material is given. Each oligomeric series

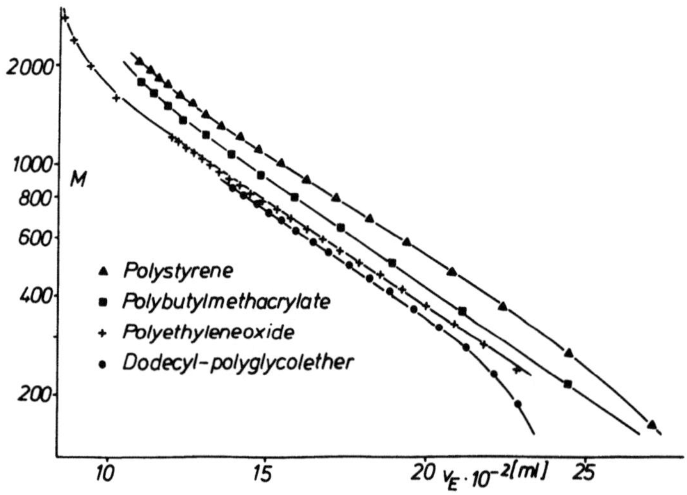

Fig. 6. GPC on polystyrene gel crosslinked with 2% divinyl benzene

gives a different calibration curve. This is due to the fact that the hydrodynamic molecular size and not molecular weight is the important factor for separation; furthermore, specific interactions between solute and gel may also shift these curves. Despite this uncertainty the basic statements still hold. Assuming that this polystyrene is used as a Merrifield resin and the activated groups starting the peptide chain are equally distributed, we see that a large portion of "pores" (> 20%) is filled up after coupling of 4 or 5 amino acid residues. This may have different consequences:

1) Some peptide chain ends are strongly restricted in the next coupling step leading to long reaction times or aborted sequences.

---

[3] A centrifugal reactor suited for handling of low crosslinked gels has been constructed by Birr (*27*).

2) As the "pores" of the gel are limited by solvated polymer chains there is no real boundary. Peptide formation can continue in neighboring "pores", forming an interpenetrating system. After cleavage of the polymer-peptide bond the extraction of the peptide is not quantitatively possible or extraction times are long.

The fact that solute size has an influence on the kinetics has been shown by Grubbs (*28*). A polymer-bound Rh-catalyst was made according to Scheme 1.

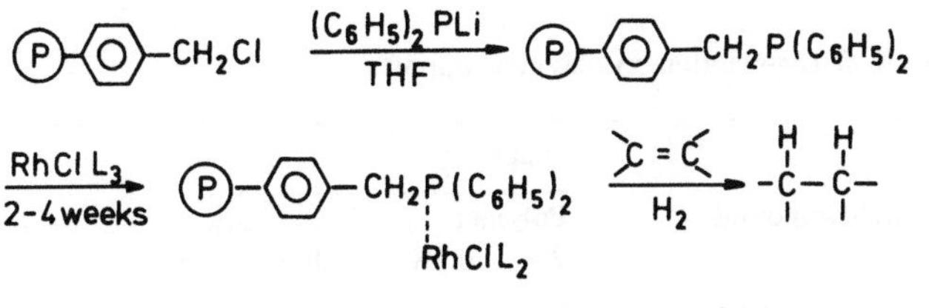

Scheme 1

Table 1. Relative rate of hydrogenation of cycloolefins using Rh Cl $[P(C_6H_5)_3]_3$ (I) and the analogous polymer bound Rh-complex (II)

Olefin	Relative rate II	I
Cyclohexene	1	1
Cyclooctene	0.39	1
Cyclododecene	0.22	0.67
2-Cholestene	0.03	0.72

Using this catalyst for the hydrogenation of cyclic olefins, the relative rate is dependent on ring size (Table 1). The rates of the low-molecular-weight reactions are not strongly influenced by ring size. Similarly, yields in Wittig olefination are dependent on the crosslinking density. Scheme 2 shows the chemistry (*29–34*).

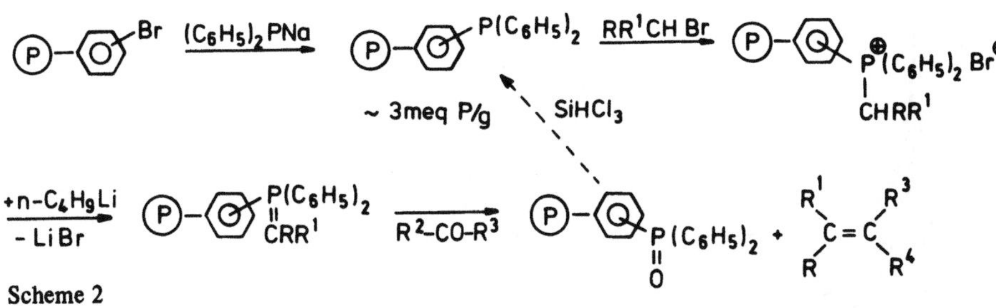

Scheme 2

A brominated polystyrene is reacted with sodium diphenylphosphan to form the polymeric Wittig reagent. Reaction with a halide and a base form the ylid which reacts with the carbonyl compound to the olefin. Wittig olefination can be made stereoselective (*33*). The formation of cis-olefins is accomplished in "salt-free solution". Using a polymeric Wittig reagent, the LiBr formed can be washed out with THF and a high cis-selectivity obtained. The reagent used can be regenerated by $SiHCl_3$.

Table 2 shows yields using polystyrenes: 2% and 0.5% crosslinked and those obtained by the conventional technique. In most cases the yields using 2% crosslinked polymer are smaller than with 0.5% crosslinked polystyrene.

Table 2. Comparison of yields in Wittig-reaction on polymer supports

RR^1CHBr	Carbonylcompound	Yield (%) Polymer 2%	0.5%	Low th. w.	lit.
Geranyl bromide	p-Chlorobenzaldehyde	11	48	49	61
	4-Benzoyl-1-bromobutane	12	26	22	24
Ethyl bromide	9-Anthracen-carbaldehyde	–	66	7	–
Isopropyl bromide	9-Anthracen-carbaldehyde	10	–	16	28
9-Fluorenyl bromide	Benzaldehyde	68	86	65	84
n-Butyl bromide	Acetophenone	100	–	100	100

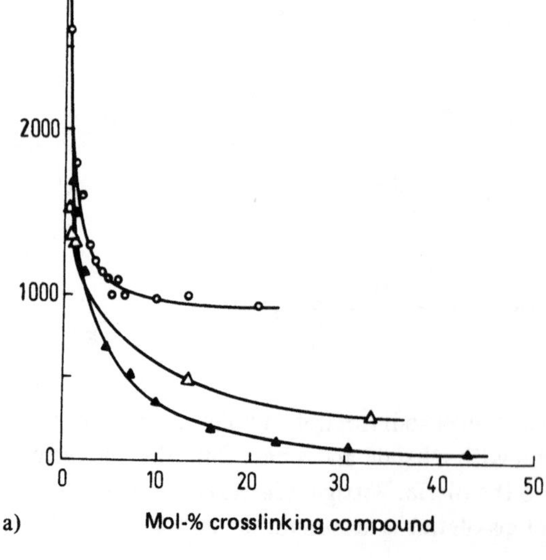

a)

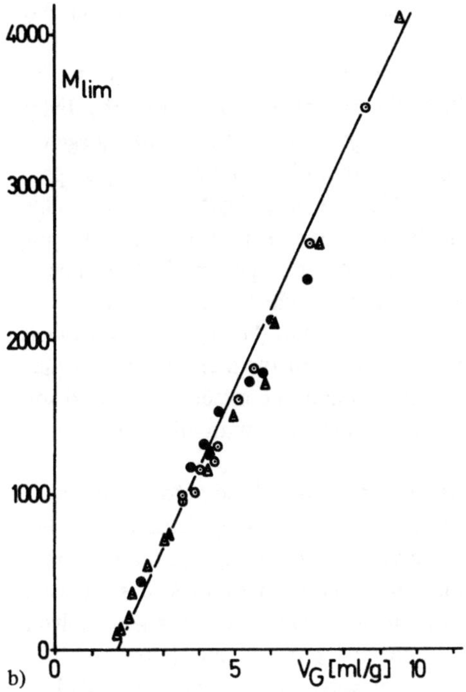

b)

Fig. 7. Excluded molecular weight of homogenously crosslinked gels obtained in THF;
a) in dependence on the amount of crosslinking agent,
b) in dependence on specific gel bed volume in THF (poured volume of swollen gel grains/g dry gel). ○ vinyl acetate – butanediol divinyl ether, △ vinyl acetate – divinyl adipate, ▲ methyl metharcrylate – ethylene glycol dimethacrylat, ● styrene-divinyl benzene

The correlation between the amount of divinyl compound and accessibility is reasonable within a series of polymers differing only in crosslinking density. Chemically different polymers polymerized with equal amounts of crosslinking agent may differ in pore structure because of the efficiency of the divinyl compound in crosslinking and a change in solvatation (Fig. 7 a). The pores are formed by swelling. Swelling data gives a better correlation with the pore structure (Fig. 7 b) (*35*). This is of special importance if the polarity of the crosslinked polymer is altered during a reaction sequence. Similarly, Birr (*36*) has explained the increasing reaction times for a complete reaction of peptides of increasing length bound to a 0.5% crosslinked polystyrene by a change from a hydrophobic to a more hydrophilic polymer.

2.3.2. Heterogenously Crosslinked Polymers

When using conventional ion exchangers, diffusion problems decrease the lower the crosslinking density. But the mechanical requirements in industrial applications force the use of higher crosslinking densities. Great progress was made when the macroporous polymers (*18, 35, 37, 38*) became available. These are extremes of a type of structure in which the crosslinking density is not randomly distributed in the polymer network.

The principal of the synthesis of heterogenously crosslinked polymers is to use a high amount of crosslinking agent and to dilute the monomer phase by an inert compound. This compound must be soluble in the monomer mixture and must be removable after polymerization. To the forming polymer, this added compound may be a solvent or nonsolvent. Even another linear polymer may be added.

Viewing a single polymerizing droplet in a suspension polymerization, we see it is surrounded by an aqueous phase and consits of the monovinyl compound, the divinyl compound, and the inert solvent. Because of the large amount of crosslinking agent and the dilution, we have an increased number of intramolecular crosslinking steps. As a result, highly crosslinked nuclei originate which are only slightly swellable. This leads to a phase separation. We now have highly crosslinked solid particles in the polymerizing droplet. These nuclei grow together in the course of polymerization. The cavities between the particles are filled with inert solvent which becomes poor in monomer. A polymer results which might be porous even in the dry state. But this is no requirement for this type of structure. Upon removal of the inert compound the porous system can collapse, forming a dense polymer and in contact with a good solvent large pores are formed again. Here again GPC provides us with information about the pore structure in the swollen state.

The most important factors governing the heterogenity of the network are the kind and the amount of inert compound as well as the crosslinking density. Very high amounts of crosslinking agent result in brittle materials, so usually one tenth to one third of the monomer mixture is divinyl compound. The proper pore size is achieved by the inert compound composed of a mixture of two solvents: one is a good solvent for the polymer, the other a precipitant. Good solvents produce small pores, bad solvents large pores. Figure 8 shows this in the case of polystyrene gels (*38*) crosslinked

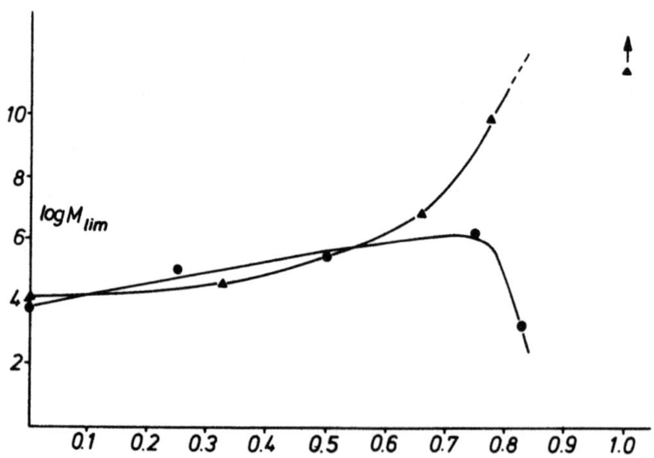

Fig. 8. Exclusion molecular weight of polystyrene gels, 30% styrene, 10% DVB, and 60% inert compound; diethylbenzene, i-amyl alcohol ▲; toluene; dodecane ●

with 25% DVB. The plotted quantities are: the exclusion molecular weight – the upper working limit in GPC – and the composition of the inert compound (volume fraction of nonsolvent). Good solvents such as toluene or diethylbenzene produce an exclusion molecular weight of about 10 000. Making the quality of the inert compound worse by use of a nonsolvent which is more polar than the growing polymer, results in a steady increase in pore size. But using a nonpolar precipitant such as a hydrocarbon, the exclusion molecular weight decreases drastically within a small change in composition (here 3 orders of magnitude). This is a common observation.

Figure 9 shows another example (*39, 40*). It is a polyvinyl acetate gel. The important fact with respect to polymeric supports is that the copolymers of the right-hand side of the maximum have the "wrong" structure. Microscopic investigations sometimes show particles shaped like a squeezed lemon; in contact with a solvent they jump up to spherical particles. Thin sections sometimes show hollow

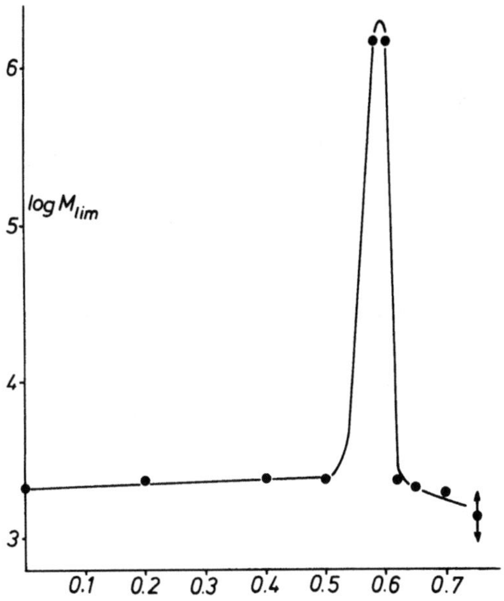

Fig. 9. Exclusion molecular weight of polyvinyl acetats crosslinked with 20% 1,4 butanediol divinyl ether; iner compound i-amyl alcohol / i-octane, abscissa given in volume fraction of nonsolvent

spheres. And even if the microscopic picture is normal, the elution behavior in GPC is characteristic (Fig. 10) (*16*). Elution of low-molecular-weight compounds let us expect a high exclusion molecular weight. But in a small molecular weight range the behavior changes and the molecules are excluded. The shape of the peaks is very informative. Very small molecules and substances totally excluded are eluted as narrow, symmetrical peaks, indicating a properly filled column. With increasing molecular size, peak-spreading grows unusually and the peaks become skewed. Because of the dense shell the solute molecular size and the pore size are similar, the diffusional restrictions are so important that a kinetically controlled behavior – a deficiency in equilibrium conditions – results.

The aim of the use of macroporous polymers is to form a polymer with a high interior surface. This surface should be so dense that we can functionalize it in a kind of monolayer. Only then are the diffusional problems largely suppressed. Usual macroporous polymers have besides the permanent porosity (porosity in the dry state) a

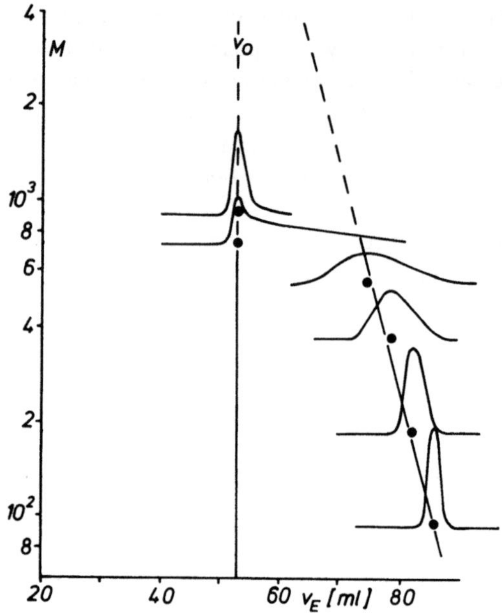

Fig. 10 Elution behavior of gels with a dense shell, PVAc with 10 wt.-% divinyl adipate (n-octane)

certain amount of swelling porosity (porosity created during swelling). Obviously the swelling porosity is contradictory to the demand for a dense surface. In order to minimize this swelling porosity techn. DVB (55%) was polymerized to get a very dense crosslinking (*41*). Figure 11 shows that using i-pentanol/toluene the exclusion

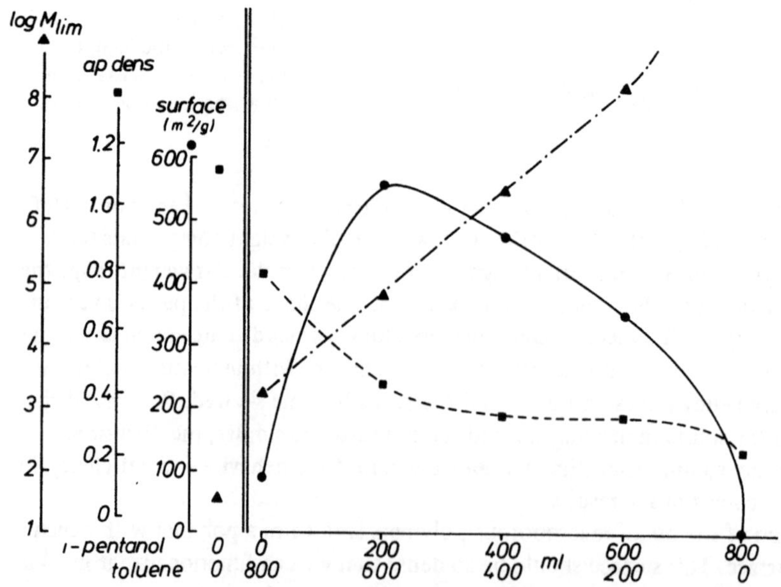

Fig. 11. Properties of gels from techn. DVB (55%, 400 ml) polymerized in presence of i-pentanol/toluene (800 ml)

molecular weight increases with decreasing solvation power. With pure pentanol the value is so high that it has no meaning physically. The interior surface is small if we have large pores and increases with decreasing pore size as long as the total pore volume is constant. Interior surfaces of 500 to 600 m^2/g can easily be obtained. The low value of the interior surface for a polymer prepared with pure toluene is due to a partial collapse of the structure during removal of the solvent. The apparent density of 0.3 to 0.4 (average density of the polymer matrix + air in the pores) is in agreement with the idea that in a highly crosslinked material the volume of added inert compound is equal to the final pore volume. But the swelling behavior is very surprising. In all cases we have a swelling of 10 to 30%, which indicates that even with 55% DVB the desired structure is not realized. To explain these and other facts we proposed the structure given in Figure 12 (*42*). The reason for its formation is the copolymerization behavior of DVB. Polymerizing a mixture of styrene and divinylbenzene to

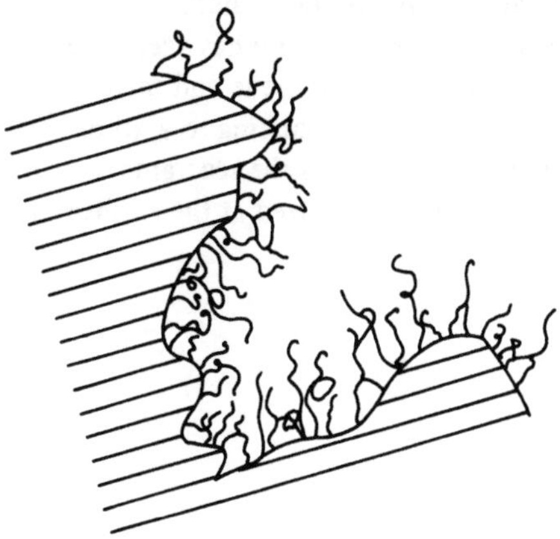

Fig. 12. Model of a macroporous gel with decreasing concentration of the "polymer solution" toward the center of the pores

low conversions the polymer will be richer in DVB than the monomer mixture. So during the last part of a polymerization of S/DVB we always have a monomer mixture rich in styrene, and the densely crosslinked microparticles are covered and connected with less crosslinked polymer. This explains the fact that on iodination or bromination the functionality is higher than expected from a surface reaction. The use of these gels for Merrifield synthesis lead to difficulties with regard to the proposed structure (*43*). GPC behavior of polystyrene gels can be explained in the same way. The elution characteristics in GPC are governed by the hydrodynamic volume of the eluted polymer molecules. As the hydrodynamic volume is represented by $[\eta] \cdot M$ ($[\eta]$-viscosity number) the plot $\log [\eta] M$ vers. ν_E will give a generalized curve holding for all polymers if pore size and pore-size distribution are the only factors responsible for separation. In the proposed structure not the whole pore is filled with pure solvent. Near the wall we have a "polymer solution" giving rise to a partition effect. This effect must obviously be more pronounced for small pores. In accordance with this, Dawkins (*44*) found increasing deviations with decreasing molecular weight from the $\log [\eta] M$ vers. ν_E calibration curve.

Summarizing macroporous systems can be done in any desired pore size range. They are of inestimable value if we deal with large molecules. In using them as polymeric supports for reactions with small molecules or a stepwise synthesis we are still faced with the problem of an appropriate surface structure.

2.4. Microgels

Another approach in the use of polymer supports for chemical reactions are the microgels. Two possibilities exist: Either a small amount of crosslinking agent can be used to form particles easily penetrated, or densely crosslinked microgels can be prepared which react at the surface. These particles can be formed by emulsion polymerization and have particle sizes between 50 and 200 Å. Slightly crosslinked gels of this type are shown to have some inhomogenity in network structure (*45*), alt although they should have reaction characteristics similar to linear polymers with the advantage that the separation by ultrafiltration is greatly facilitated.

The formation and reactivity of microgels of a dense structure has been investigated by Funke and Seitz (*46–53*). By use of pure p-divinylbenzene and techn. DVB (55%), respectively, the polymers have a vinyl content corresponding to a 40 to 90% consumption of the second vinyl group of DVB. Depending on the reactants and the reaction conditions subsequent reactions may be localized at the surface of the particle.

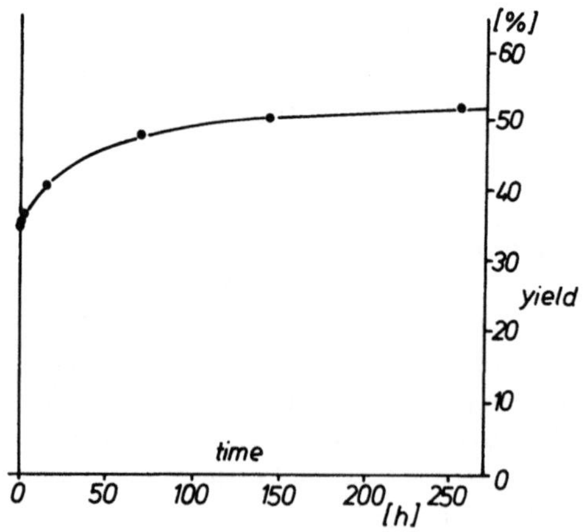

Fig. 13. Reaction of double bonds in microgels with Hg Ac_2

Figure 13 shows the reaction of the double bond with Mercury (II) acetate (*50*). In a fast reaction the double bonds at the surface react. Subsequently controlled by a slow diffusion process, the double bonds in the interior of the particle react. The specific surface of these systems is 200 to 300 m^2/g. As enzyme carriers they should be well-suited, as support for polymeric reagents more knowledge about the possibility of localizing the reaction at the surface is needed.

3. The Microstructure

The number of publications on polymeric reagents has been rapidly growing during the last five years. Reviews were published by Overberger (*9*) and Leznoff (*10*). I would like to mention just some points which seem to be useful guidelines for application.

Fixation of a reactant permits work under high-dilution or concentrated-solution conditions.

As Patchornik (*54, 55*) showed the competitive self-condensation of an ester can be avoided if the ester molecules are separated by fixation on the polymer. According to Scheme 3, 0.1 to 0.2 meq ester groups were fixed at the chloromethylated polymer. Pure monoalkylation can be achieved and no self-condensation is observed if the load is kept small.

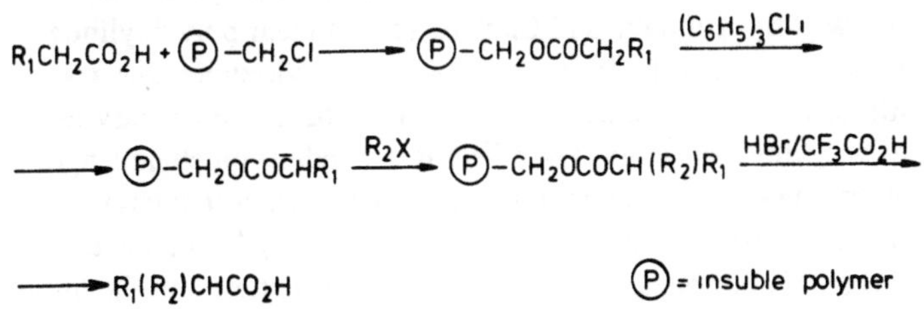

Scheme 3

The opposite situation is given if two molecules fixed at the polymer should react. According to Scheme 4 a small amount of the chloromethyl groups are reacted with an enolizable acid the rest with an nonenolizable acid. Addition of a base results in the formation of only one ketone, without the use of a polymeric support a complex mixture results (Table 3) (*56*).

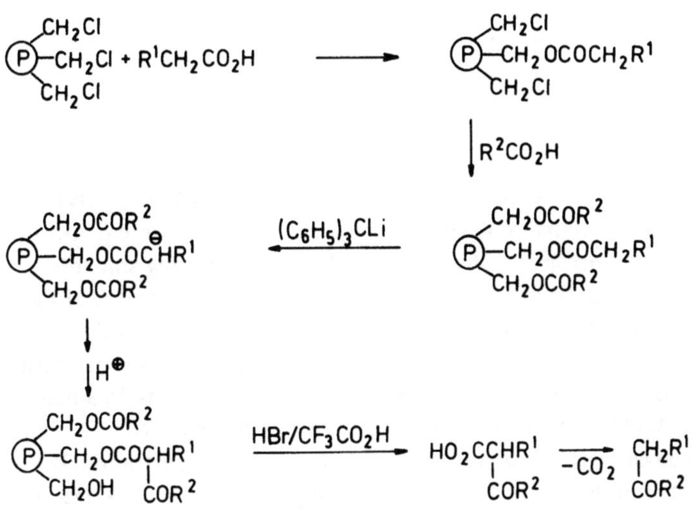

Scheme 4

Table 3. Alkylation of polymer bound esters

Ester	Alkylating agent	Product	Yield (%)
P $-CH_2-COO-CH_2-C_6H_5$	CH_3-	$C_6H_5-CH(CH_3)-COOH$	85
P $-CH_2-COO-CH_2-C_6H_5$	$C_6H_5CH_2-$	$C_6H_5-CH(CH_3)-COOH$	50
P $-CH_2-COO-CH_2-CH_2-C_6H_5$	$n-C_4H_9-$	$C_6H_5-CH_2-CH(C_4H_9)-COOH$	45

In aqueous solution the catalytic esterolyses with polymeric catalysts is accelerated by "hydrophobic bonding". The catalytic effect of synthetic polymers on hydrolysis has been reviewed by Kunitake and Okahata (*57*). Linear polyvinylimidazole (Scheme 5) increases the rate of hydrolyses of 3-nitro-4-acyloxybenzoic acid by a factor of 10^3 with respect to monomeric imidazole (*58–62*). The primary reaction is the acylation of polyvinylimidazole. The effect is dependent on chain length of the acid and the solvent. The hydrophobic interaction causes an autocatalytic course of reaction rate. The acylated polyvinylimidazole is more hydrophobic than the starting polymer. At a 75% conversion the rate is five times higher than the initial rate.

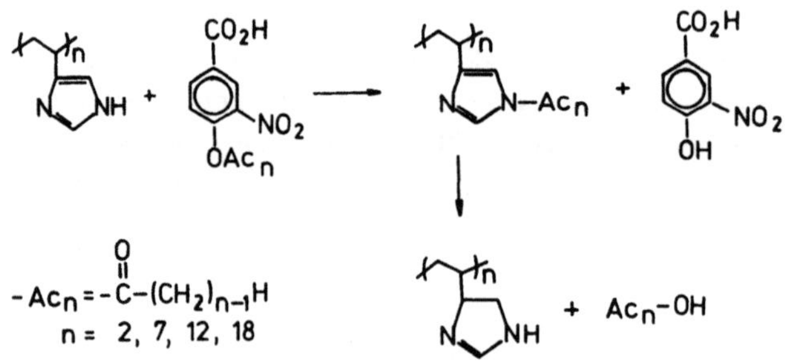

Scheme 5

Similar hydrolysis studies were made by Morawetz (*63*) on crosslinked polyacrylamide carrying pendent pyridine residues (Table 4). The substrates were p-nitrophenylacetate (NPA) and p-nitrophenyl-butyrate (NPB). With increasing amounts of crosslinking agent the catalytic efficiency on hydrolyses increases. Because of the lower swelling ratio the swollen polymer becomes increasingly dissimilar to the surrounding medium and the local concentration of pyridine groups increases. But with higher amounts than about 2% of crosslinking agent diffusional problems dominate. So the hydrolysis rate shows a maximum at medium crosslinking densities.

One of the main advantages of polymeric reagents is the ease of separation. But they can also solve other problems. In the preparation of polymeric peracids one of

Table 4. Catalytic activity on hydrolysis; crosslinked acrylamide with pendant pyridine residues. DC-degree of crosslinking

DC (%)	Swelling ratio	Catalytic efficiency	
		NPA	NPB
0.5	17.0	0.33	0.59
1	12.4	0.74	1.5
2	7.1	1.00	4.5
3	5.6	0.66	2.8
4	5.0	0.21	0.91

the desirable successes would be to increase safety. On the other hand, a high capacity is essential for use. Derived from a crosslinked copolystyrene-maleic anhydride a polymeric peracid was prepared with up to 7 meq $-C(=O)OOH$ /g, which showed to be quite stable when immersed in a solvent and gave high yield in subsequent reactions. But on putting it into another bottle in the dry state, the electrostatic charge at the surface initiated an explosion (*64*). Resins containing aromatic per-acid residues are insensitive to shock (*65*).

Polymeric reagents are also of advantage if the corresponding low-molecular-weight reagent has an extreme odor or high toxicity. Selenophenol is such an example. Polymeric selenium can be prepared in three ways according to Scheme 6 (*14, 66*). Scheme 7 shows examples of their use. Reaction with a ketone and cleavage of the adduct with hydrogen peroxide yield an unsaturated ketone. α-halocarboxylic esters form the unsaturated esters in high yields. Activated methyl groups can be selectively converted to the aldehyd without a trace of the corresponding acid. Olefins form diols.

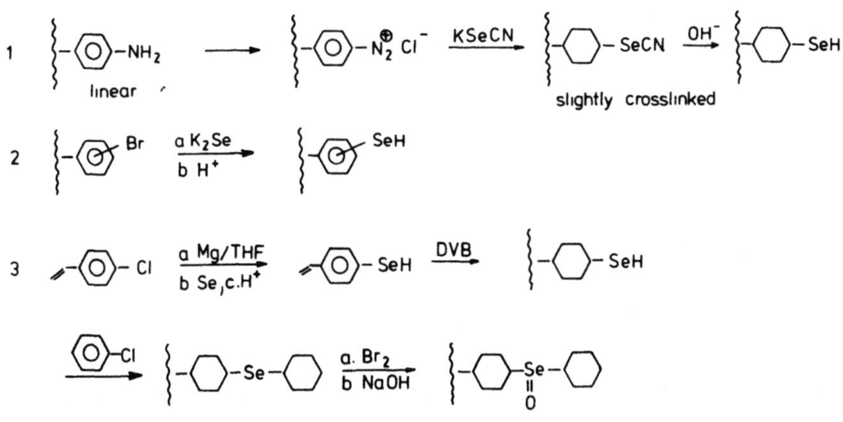

Scheme 6

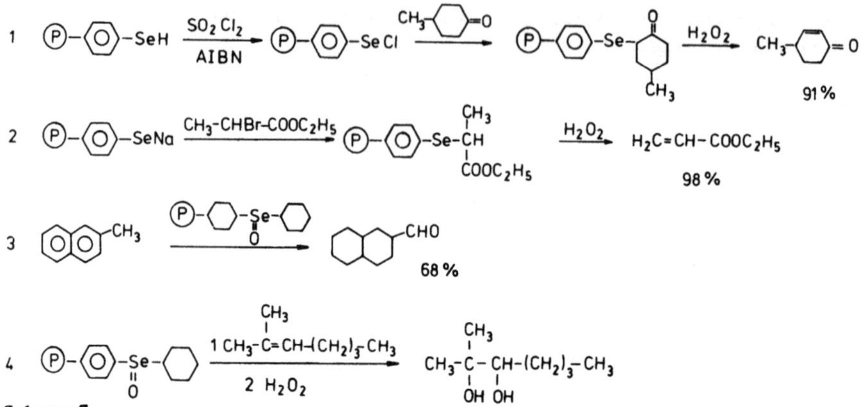

Scheme 7

Polymeric supports can also be used with advantage to form monofunctional moieties from difunctional ones. Leznoff has used this principal in the synthesis of sex attractants on polymer supports (*67*). Starting from a sheap symmetrical diol he blocked one hydroxyl group by the polymer. Functionalization of crosslinked polymers is mostly performed by chloromethylation (*68*). A very promising method to introduce functional groups into crosslinked styrene-divinylbenzene copolymers is the direct lithiation with butyllithium in presence of N,N,N′,N′-tetramethylethylenediamine (TMEDA) (*69, 70*). Metalation of linear polystyrene with butyllithium/TMEDA showed no exchange of benzylic hydrogen and a ratio of attack at m/p-position of 2:1 (*71*). In the model reaction of cumene with amylsodium, a kinetic control of the reaction path is established. After 3h of treatment with amylsodium, cumene is metalated 42% in α-, 39% m-, and 19% p-position. After 20h the mixture equilibrates to affort 100% of the thermodynamically more stable α-product (*72*).

A field which opens new possibilities in the mechanisms of reactions is the use of solid phase chemistry for the detection of reactive intermediates. Figure 14 shows

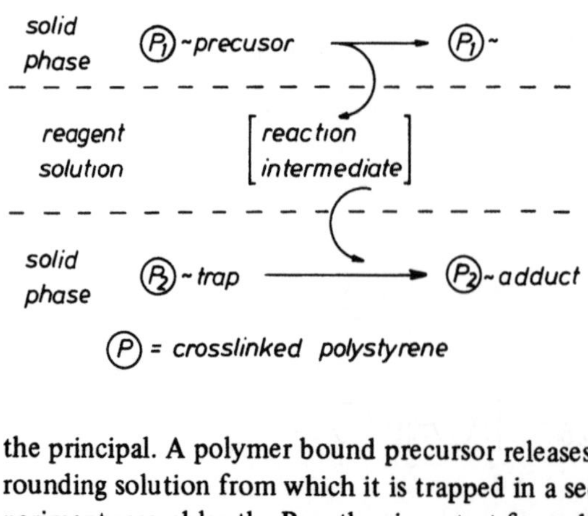

Fig. 14. Trapping of reactive intermediates using polymeric reagents (*74*)

the principal. A polymer bound precursor releases a reactive intermediate to the surrounding solution from which it is trapped in a second polymer. This type of experiment resembles the Paneth mirror test for radicals.

Aminolysis of nitrophenolesters is catalyzed by imidazole. If amino and ester groups are bound to different polymers, no reaction occurs in the absence of imidazole. Catalytic amounts of imidazole result in an acylation of the aminopolymer (Scheme 8) (*73*). So an intermediate must be formed transporting the acyl moiety from one polymer particle to the other; presumably this is the acyl imidazole.

The iron phenanthroline derivative in Scheme 9 is a source of cyclobutadiene which can be trapped by maleanil. Both reactants bound to a polymer give, when oxidized, the polymeric adduct shown. By methaminolysis a derivative is formed, proving that cyclobutadiene is a free-existing intermediate in this reaction (*74*).

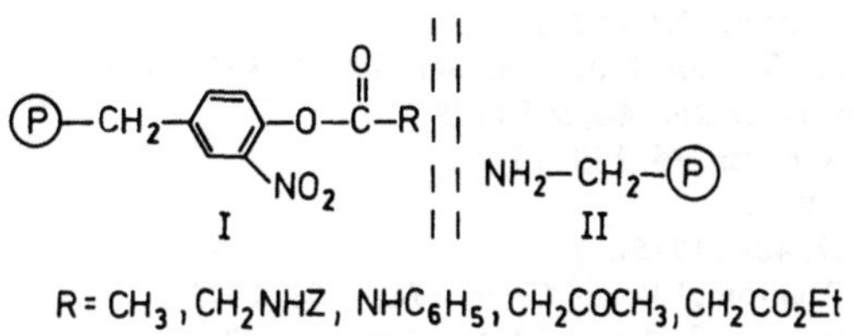

Scheme 8

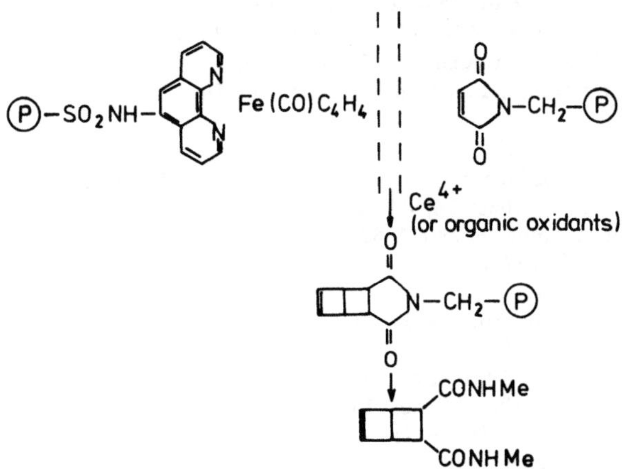

Scheme 9

4. References

1. Adam, B. A., Holmes, E. L.: J. Soc. Chem. Ind. (London) **54,** 1.
2. D'Alelio, G. F.: (to General Electric Co.) U. S. Pat. 2366 007 (1944).
3. Merrifield, R. B.: J. Am. Chem. Soc. **85,** 2149 (1963).
4. Cassidy, H. G.: J. Am. Chem. Soc. **71,** 402 (1949);
Updegraf, I. H., Cassidy, H. G.: J. Am. Chem. Soc. **71,** 407 (1949);
Cassidy, H. G., in: Encyclop. Polymer Sci. Technol. Vol. 5, 735 (1966).
5. Manecke, G.: Z. Elektrochem. **57,** 189 (1953);
Manecke, G.: Pure and Appl. Chem. **38,** 181 (1974).
6. Goldman, R., Goldstein, L., Katchalsky, E. in: Biochemical aspects of reactions in solid supports, Stark, G. R. (ed.). London: Academic Press 1971.
7. Campbell, D. H., Leuscher, E., Lerman, L. S.: Proc. Nat. Acad. Sci. U.S. **37,** 575 (1951).
8. Grubhofer, H., Schleith, L. Z.: Naturwissenschaften **40,** 508 (1953).
9. Overberger, C. G., Sannes, K. N.: Angew. Chem. **86,** 139 (1974).
10. Leznoff, C. C.: Chem. Soc. Rev. **3,** 65 (1974).
11. Bayer, E., Schurig, V.: Angew. Chem. **87,** 484 (1975).
12. Brandstetter, F., Schott, H., Bayer, E.: Makromolekulare Chem. **176,** 2163 (1975).
13. Kern, W., Schulz, R. C. in: Houben-Weyl 14, 2. Stuttgart: Thieme 1963, p. 637.
14. Michels, R., Kato, M., Heitz, W.: Makromolekulare Chem., in press.
15. Schuttenberg, H., Klump, G., Kaczmar, U., Turner, S. R., Schulz, R. C.: J. Macromolecular
16. Heitz, W., Platt, K. L.: Makromolekulare Chem. **127,** 113 (1969).
17. Brunauer, S.: Adsorption of gased and vapors. Princeton Univ. Press 1943.
18. Seidl, J., Malinsky, J., Dusek, K., Heitz, W.: Adv. Polymer Sci. **5,** 114 (1967).
19. Ritter, H. L., Drake, L. C.: Ind. Eng. Chem. Anal. Ed. **17,** 786 (1945).
20. Altgelt, K. H., Segal, L.: Gel Permeation Chromatography, New York: Dekker 1971.
21. Kubin, M.: J. Chromatography.
22. Heitz, W.: Z. Anal. Chem. **277,** 323 (1975).
23. Van Deemter, J. J., Zuiderweg, F. J., Klinkenberg, A.: Chem. Eng. Sci. **5,** 271 (1956).
24. Giddings, J. C., Mallik, K. L.: Analytic. Chem. **38,** 997 (1966).
25. Heitz, W., Coupek, J.: J. Chromatography **36,** 290 (1968).
26. Regen, S. L.: Macromolecules **8,** 689 (1975).
27. Birr, Ch., Lochinger, W.: Synthesis **1971,** 319.
28. Grubbs, R. H., Croll, L. C.: J. Am. Chem. Soc. **93,** 3062 (1971).
29. Heitz, W., Michels, R.: DOS 2052813 (1970).
30. Camp, F., Castells, J., Font, J., Vela, F.: Tetrahedron Lett. **1971,** 1715.
31. McKinley, S. V., Rahshys, J. W.: J. Soc. Chem. Comm. **1972,** 134.
32. Heitz, W., Michels, R.: Angew. Chem. **84,** 296 (1972).
33. Heitz, W., Michels, R.: Liebigs Ann. Chem. **1973,** 227.
34. Michels, R., Heitz, W.: Makromolekulare Chem. **176,** 245 (1975).
35. Heitz, W.: J. Chromatography **53,** 37 (1970).
36. Birr, Ch., priv. comm.
37. Corte, A.: Ger. Pat. 1021 166 (1957).
38. Moore, J. C.: J. Polymer Sci. **A2,** 835 (1964).
39. Bier, P.: Dissertation, Mainz 1973.
40. Heitz, W., Bier, P.: Makromolekulare Chem. **176,** 657 (1975).
41. Heitz, W., Michels, R.: Makromolekulare Chem. **148,** 9 (1971).
42. Heitz, W., Kern, W.: Angew. Makromolekulare Chem. **1,** 150 (1967).
43. Merrifield, R. B.: Advan. Enzymol. **32,** 221 (1969).
44. Dawkins, J. W., Hemming, M.: Makromolekulare Chem. **176,** 1777, 1795, 1815 (1975).
45. Hoffmann, M.: Makromolekulare Chem. **175,** 613 (1974).
46. Kühnle, D., Funke, W.: Makromolekulare Chem. **139,** 255 (1970).
47. Kühnle, D., Funke, W.: Makromolekulare Chem. **158,** 135 (1972).
48. Beer, W., Kühnle, D., Funke, W.: Angew. Makromolekulare Chem. **23,** 205 (1972).

49. Obrecht, W., Seitz, U., Funke, W.: Makromolukulare Chem. **175**, 3587 (1974).
50. Funke, W., Beer, W., Seitz, U.: Progr. Colloid a. Polymer Sci. **57**, 48 (1975).
51. Obrecht, W., Seitz, U., Funke, W.: Makromolekulare Chem. **176**, 2771 (1975).
52. Seitz, U.: Deutsche Farbenzeitschrift **29**, 450 (1975).
53. Obrecht, W., Seitz, U., Funke, W.: Makromolekulare Chem., in press.
54. Patchornik, A., Kraus, M.: J. Am. Chem. Soc. **92**, 7587 (1970).
55. Kraus, M., Patchornik, A.: Isr. J. Chem. **9**, 269 (1971).
56. Kraus, M., Patchornik, A.: J. Am. Chem. Soc. **93**, 7325 (1971).
57. Kunitake, T., Okahata, Y.: Adv. Polymer Sci. **20**, 159 (1976).
58. Overberger, C. G., Morimoto, M., Cho, J., Salomone, J. C.: Macromolecules **2**, 553 (1969).
59. Overberger, C. G., Glowaky, R. C.: J. Amer. Chem. Soc. **95**, 6014 (1973).
60. Overberger, C. G., Morimoto, M.: J. Amer. Chem. Soc. **93**, 3222 (1971).
61. Overberger, C. G., Morimoto, M., Cho, J., Salomone, J. C.: J. Amer. Chem. Soc. **93**, 3228 (1961).
62. Imanishi, Y., Amimoto, Y., Sugihara, T., Higashimura, T.: Makromolekulare Chem. **177**, 1401 (1976).
63. Morawetz, H.: Pure a. appl. Chem. **38**, 267 (1974).
64. Tauchert, K.: Dissertation, Mainz 1974.
65a) Hodge, P., Kemp, J., Perry, G. M.: Makromolekulare Chem. **176**, 267 (1975).
b) Fréchet, J., Haque, K. E.: Macromolecules **8**, 130 (1975).
66. Kato, M., Michels, R., Heitz, W.: J. Polymer Sci. B, Polymer Letters, in press.
67. Wong, J. Y., Leznoff, C. C.: Can. J. Chem. **51**, 2452 (1973).
68. Patterson, J. A. in: Biochemical aspects of reactions in solid supports. Stark, G. R. (ed.). London: Academic Press 1971.
69. Leznoff, C. C.: Proc. of "table ronde roussel uclaf" on Polymer Bound Molecules, Paris 4./5. March 1976.
70. Hartmann, H.: Makromolekulare Chem., in press.
71. Evans, D. C., Phillips, L., Barric, J. A., George, M. H.: J. Polymer Sci. Polymer Lett. Ed. **12**, 199 (1974).
72. Benkeser, R. A., Liston, T. V.: J. Amer. Chem. Soc. **82**, 3221 (1960).
73. Rebek, J., Brown, D., Zimmermann, S.: J. Amer. Chem. Soc. **97**, 454 (1975).
74. Rebek, J., Gaviña, F.: J. Amer. Chem. Soc. **96**, 7112 (1974).

Received September 27, 1976

Polymeric Drugs

Hans-Georg Batz*
Institut für Organische Chemie der Universität Mainz

Table of Contents

---

* New address: Boehringer Mannheim GmbH, Biochemica, Werk Tutzing, Bahnhofstr. 5, 8132 Tutzing/Obb., Germany

1. Introduction

Nowadays polymeric substances have attained invaluable importance in medicine as well as in nearly every sphere of life. Their relatively easy variability of physical and chemical properties enables accomodation to many different requirements. Examination of new monomeric compounds and the combination of known monomers to new copolymers continue to enlarge the palette of the various substances. In spite of the fact that a macromolecular product was already introduced in 1894 into the medicinal practice by Fränkel, when he implanted celluloid into the skull-cap, it took decades until the importance of the implantation of artificial material on the basis of synthetic polymers was completely recognized. The application to the human organism was only possible due to the enormous scientifique and technological development in the polymeric field. The basic condition necessary for artificial material used for implantations is its absolute inertness towards biological systems. The development and the importance of artificial material for implantations and of the applied substances has been discussed in several reviews (*1–5*). Today the general interest is directed towards the toxicity of traces of monomers and other substances of low molecular weight present in the polymeric material (*6*) as well as towards the compatibility of polymer surfaces with the blood and the tissue. The great number of publications concerned with these problems (*7*) show that, so far, no satisfying solution has been found.

Only during the last two decades has it been recognized that the interaction of synthetic polymers with biological material can also be favorable and desirable. Thus, polyglycolide and polylactide, which are applied as suture materials in surgery, are slowly degraded and resorbed in the body (*8*). Definite pharmacological properties have been discovered in the case of other polymers, *e.g.* the antisilicosis activity of poly(vinylpyridine oxide) (*9*) or the interferon induction by copolymers (*10*) from maleic anhydride and divinyl ether. These recent developments reduced the scepticism still partially present towards the in vivo application of synthetic not-inert polymers and inspired a great number of new investigations. The first studies date back to the fifties (*11*). Several review articles (*10–14*) show, that most of the studies originate from Russian groups like that of Ushakov in Leningrad, where syntheses and biological activities of polymeric drugs had already been investigated at that time (*17*).

2. Definitions

The term "polymeric drugs" comprises not only pharmacologically active polymers, the monomeric derivatives of which are inactive, but also substances in which a known drug is covalently bound to a polymeric carrier or is incorporated into the polymeric chain itself. The average molecular weight must not pass a certain value (as a rule[1] 10 000 to 30 000); the polymeric molecules can be composed of one or several repeating units. Unlike low molecular compounds and several naturally occurring compounds of high molecular weight, synthetic polymers consist of a mixture of molecules of different chain length.

Based on their chemical structure polymeric drugs are divided into the following three groups:

a) Polymers and copolymers which are active only as macromolecules.
b) Drugs in which the active substance of known biological activity is bound to a polymeric carrier molecule via a functional group.
c) Drugs, which are obtained from an active substance, containing in the sense of macromolecular chemistry at least two functional groups, by polycondensation with itself and eventually with another bifunctional component.

3. Polymers and Copolymers Exclusively Active as Macromolecules

Although naturally occurring macromolecules like DNS, proteins and carbohydrates belong to the basic components of biological material, synthetic macromolecules and their interactions with biological systems escaped notice for a long time. This may be due to the fact that the knowledge on the structure, the interactions, and the biological importance of naturally occurring macromolecules has only been substantially improved during the last decades and, on the other hand, appropriate statements on synthetic polymers are often considerably more complicated or not even as yet possible, since in this case, particularly, substances of inhomogeneous molecular size and structure are concerned. The activity of synthetic polymers like poly(2-vinylpyridine) and poly(4-vinylpyridine) against silicosis, for the first time described in 1960 by Schlipköter and Brockhaus (*18*), initiated a great number of investigations at first in the research of silicosis but recently also in other areas like cancer research.

3.1. Polymers with Antisilicosis Activity

Since the applied poly(vinylpyridine)s are only soluble in form of their salts in acidic solution and, therefore, precipitate in the organism of the animal after injection, the application of the corresponding N-oxides, like poly(2-vinylpyridine-1-oxide) PVNO [*1*] constitutes a decisive improvement (*19*).

By subcutaneous or intravenous injection of an aqueous solution of (PVNO) (*20*) or by inhalation of an aerosol of a PVNO-solution (*21*) the cytotoxical action of quartz dust can be prevented prophylactically and acute or chronic silicosis can be

[1] Since the rate of excretion represents a substantial criterion for polymeric drugs and since this rate can be retarded in the case of molecular weights, which are still higher, such a value may be supposed, considering the fact that the molecular weight is, however, not always the sole determinant factor.

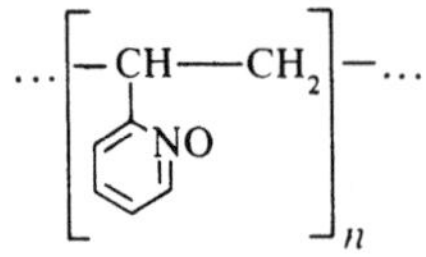 *1*

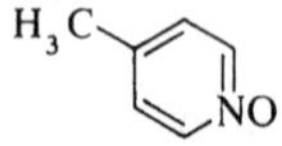 *2*

treated therapeutically. It has been shown that 4-methylpyridine-1-oxide [*2*], an analogous low molecular model compound of *1*, is completely inactive (*20*), and that an optimal activity of polymer *1* is attained only in the case of molecular weights > 30 000.

The adsorption of PVNO of low basicity at the surface of quartz covered with a layer of silicic acid is referred to as one of the reasons for their activity (*21*).

The neutralization and the complexation with PVNO of this surface of quartz is of decisive importance (*22*) since today the proteolytic enzymes of the lyosomes are supposed to dissolve the layer of proteins and other body-specific substances on the quartz particles, and thereby expose the structure of the crystal lattice of the SiO_2-particles, which then in direct contact with the cell components cause the damage of the cell mechanism leading at last to the death of the macrophages (*22*).

Based on these assumptions of the mechanism of action, Ferrutti and Marchisio (*23*) have examined several N-oxides like poly(3-vinylpyridin-1-oxide), poly(N-allyl-piperidine-N-oxide), poly(*p*-dimethyl-aminostyrene-N-oxide) and other analogous substances. They have been able to demonstrate that these substances show the same activity.

Holt, Lindsay and Beck (*24*) have studied the type and strength of adsorption of the polymers at the surface of quartz. They showed that the configuration of PVNO (isotactic or syndiotactic) has some influence on the power of adsorption and also on the biological activity.

Poly(vinylpyridinium) salts, however, are only active if the quartz particle had been treated with them before its action. Poly(vinylimidazole) (*25*) is inactive, although it is as strongly adsorbed as PVNO.

These and other results force one to the assumption that the activity of PVNO and other polymeric N-oxides against silicosis must be based not only on the adsorption of the dust particles but also on some interactions with the biological system of the organism. As early as 1963 Antweiler (*26*) was able to show that PVNO, in contrast to poly(vinylpyrrolidone), and the homologues of low molecular weight [like *2*], accelerate greatly the rate of sedimentation of human and animal erythrocytes and increases considerably the resistance of erythrocytes against hypotonic solutions of sodium chloride. Manojlovic (*27*) has studied the influence of PVNO on the growth of cells of "line L"[2] and has found a change in the frequency and velocity of mitosis. In addition an increased death rate by the influence of PVNO could be established. In a review article Holt (*28*) has discussed additional effects on biological systems like enzymes: So far a conclusive statement on the mechanism of action seems, however, not to be possible.

It should still be mentioned, that Papilian and co-workers (*37*) have found some activity of Triton ®-like substances against silicosis.

[2] Cells of "line L" are cells of a mouse cell line in a cell culture.

3.2. Polymers with Metastases-Stunting Properties

The widely accepted opinion, which supposes that not only an adsorption at the quartz particle but also an interaction with the cell membrane occurs, may have been the reason for Ferrutti and co-workers (*29*) to study the effect of such polymeric N-oxides on the cancer metastasis. Previously it has been shown by Franchi and co-workers (*30*), that polymeric detergents of the Triton ® type [*3*] show strong metastases stunting activity.

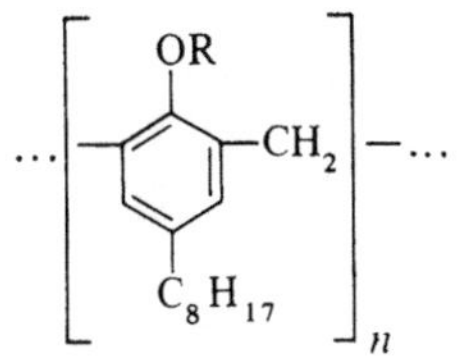

3

$R = -CH_2CH_2O-(CH_2CH_2O)_m-CH_2CH_2OH$

$m = 1-10$

$n = 4-10$

Since n and m are small numbers, these drugs in question are, in the proper sense, no polymers, but defined oligomers with a uniform molecular weight. The preparation *3* does not show any activity on the transplanted primary tumor (Ehrlich Carcinom and Sarkom 180), however, stunts very strongly the formation of metastases, if applied parenterally between the 7th and 20th day. As far as the mechanism of action is concerned, on the one hand an activation of the RHS-system[3] is discussed and on the other hand an interaction with the membrane of the cancer cells supplied by the primary tumor into the blood-vessels is supposed.

By such an adsorption of *3* at the membrane it is intended to prevent an establishment of cancer cells in the healthy tissue. A somewhat different interpretation of the action of such detergents has been given by Calvin (*31*), who supposes that the enzymes are extracted from the cell membrane by the formation of detergent micels or that inhibitors are stabilized in such a way that they cannot interact anymore with the enzyme.

Similar surface active but polymeric substances, partly possessing additional cationic properties [*e.g.* *4*], have been synthesized by Ferrutti *et al.* (*29*).

Besides such poly(amide-amine)s a poly(acrylamide) derivative of structure *5* was also active.

---

[3] The part of organism, which is of substantial importance for the kind of action of polymeric drugs is called the RHS- and RES-system (reticulo-histiocyteric and reticulo-endothelial system). *E.g.* the reticulum and endothel cells of the spleen and of the lymphatic gland as well as the "von Kupffer star cells". These cells among other things are capable of taking up macromolecular substances which cannot be transported or diffused through the cell membrane by phagocytosis or pinocytosis, *i.e.* by inclusion of the macromolecular substance in a dent of the cell membrane followed by a ligature in the interior of the cell. The macromolecular substances are deposited in the interior of the cell and they can be delivered to the blood by an appropriate mechanism. Finally, the formation of antibodies, initiated by macromolecular substances, has to be localized in several of these cells.

The interferon induction with similar pharmacological effects such as immunization is also connected with an activation of these cells.

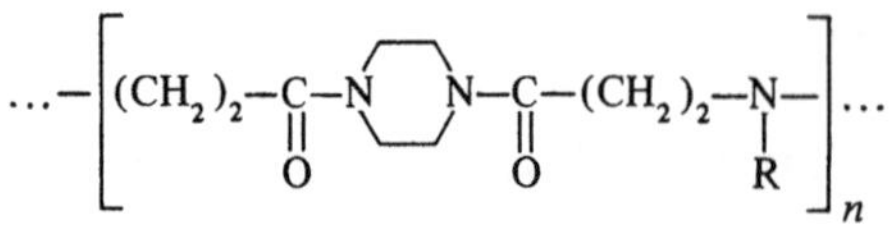

R = $(CH_2)_{11}CH_3$ (10–30 mol-%)
and $(CH_2)_2OH$ (70–90 mol-%)

4

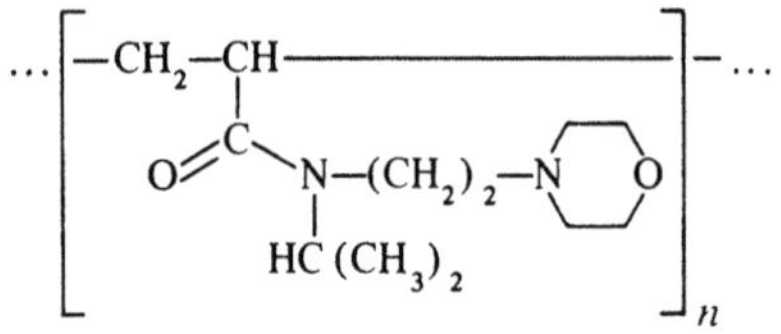

5

Polymeric N-oxides, like *1*, show no activity in this test. An activation of the RHS-system has also been supposed to be the reason for the antitumor action of divinyl ether/maleic anhydride copolymers [*8*] which have been described for the first time by Regelson (*32*) as interferon inductors.

3.3. Other Surface Active Substances with Pharmacological Activity

Already in 1951 Cornforth and co-workers (*33*) have shown that Triton® [*3*] analogous substances are active in experimental tuberculosis. The same team (*34*) has recently published several similar new preparations containing poly(ethylene oxide) residues with 10 to 60 ethylene oxide monomeric units. The length of this polymeric chain is not of importance as far as the antituberculosis activity is concerned, but in respect to the proportion between hydrophilic and hydrophobic components in the molecule. An enlargement of the poly(ethylene) chain has obviously no influence on the toxicity of the substance. Dittmann (*35*) obtained a similar result in his investigation on the local and endoanaesthetic activity of alkyl-poly-glykol ethers of the general formula *6*

$$C_nH_{2n+1}-(OCH_2CH_2-)_mOH$$

6

Variation of n between 1 and 14 and of m between 2 and 14.

Finally, Weiner and Zilkha (*36*) have investigated the prolongation of the local anaesthetic action of procain, bound to tetraethylene glykol or poly(ethylene glykol) 400[4] [7]. The tetraethyleneglykol derivative showed an extended and stronger activity.

[4] Product of Chem. Werke Hüls; average degree of polymerization $n \approx 9$.

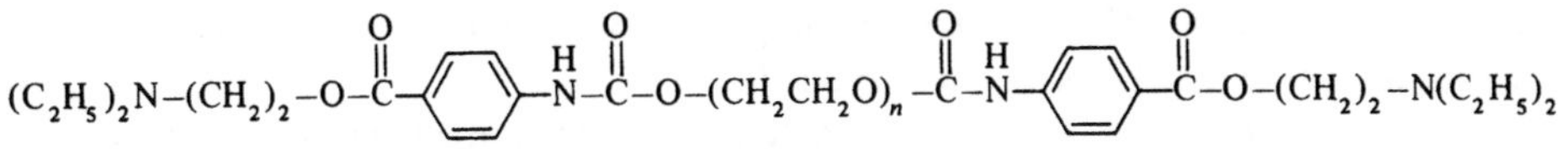

7 $n = 4$ or 9

3.4. Polymers with Antiviral Properties

Not only PVNO [*1*] but also the so-called pyrane copolymer stimulated the study of the physiological properties of synthetic polymers. This cyclic copolymers [*8*] from divinyl ether and maleic anhydride, first time described by Regelson (*38*), induces the formation of interferon.

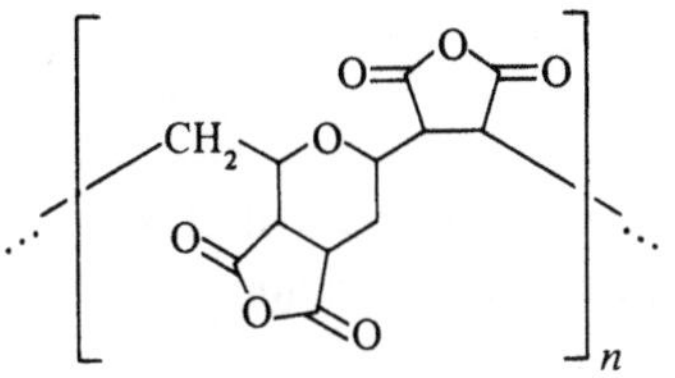

8

Since interferon has very great importance in the virus and cancer therapy forty publications have appeared on this substance in the sequel. Recently, Breslow *et al.* (*39*) have been able to show that the toxicity of the copolymer increases with increasing molecular weight. However, a molecular weight of at least 3000 to 5000 is necessary to show some activity. Copolymers of lower toxicity having the same activity could be obtained by preparation of products with a smaller molecular weight distribution, since in these products parts of higher molecular weights are not present.

Due to the toxicity of the co-(cyclopolymer) *8*, the capability of interferon induction of a great number of polymers has been tested. Thus, interferon is formed by the presence of natural double helix RNA, which occurs in some viruses (Reoviruses), but also by the synthetically produced complex (*40*) from poly(inosinic acid) and poly(cytidylic acid) (1:1). An improved and extended interferon induction of such poly(nucleic acid)s can be achieved by complexing with synthetic polycations, as for example diethylaminodextrane (*41*) or with a polycation (*42*) of composition *9*.

$$(CH_3)_2N{-}(CH_2)_n{-}\left[{-}\overset{CH_3}{\underset{CH_3}{N^+}}{-}(CH_2)_m{-}\overset{CH_3}{\underset{CH_3}{N^+}}{-}(CH_2)_n{-}\right]_x{-}N(CH_3)_2 \qquad 9$$

$2 \leqslant n \leqslant 10 \qquad 2 \leqslant m \leqslant 10 \qquad x \geqslant 2$

Since the stability of such complexes towards an enzymatic attack seems to be of great importance as far as their activity and duration of activity is concerned, the activity of synthetic poly(vinylpurin)s and poly(vinylpyrimidine)s, which cannot be degraded enzymatically, has been examined.

Pitha and co-workers (*43*) have been able to demonstrate in a series of studies the poly(vinyl uracil) [*10*] and poly(vinyladenin) [*11*] to show in fact antiviral activity, but probably no interferon induction. Rather, it has been shown, that these electrically neutral polymers with a molecular weight

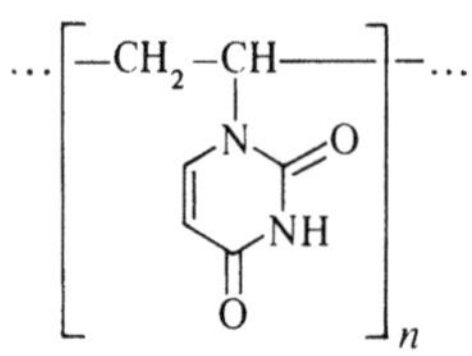

10

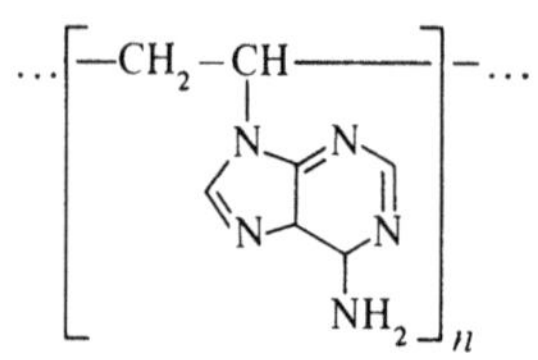

11

> 100 000 and which are resistant towards chemical and encymatic hydrolysis, are resorbed by the cells via pinocytosis and then engage in the duplication of DNS or block the protein biosynthesis (*44*), respectively.

It seems, however, that only polyanions are chiefly concerned in the interferon induction. Further polymers having anionic groups have been investigated besides poly(nucleic acid)s and co-(cyclopolymers) *8*, for which a partial hydrolysis of the anhydride moiety is supposed. Thus, poly(vinylsulfonate)s or polyacrylates have been found to be active, but, polymethacrylates inactive. It has been shown that in the case of acrylic acid copolymers the molecular weight is also of importance in addition to the structure of the polymer chain and the fraction of carboxy groups. The influence of these various parameters has been discussed in detail by De Somer (*45*).

Actually it is doubtful that the antiviral properties of polyanions are based only on the interferon induction (*46*), since several groups have also demonstrated other activities of polyanions. Thus, *e.g.* poly(vinylsulfonate) has been found to interact with nucleic acids, thus inhibiting their acylation or methylation (*47*). Tempel (*48*) has shown that in mice phosphatases can be activated with poly(vinylsulfonate).

Since many of such detailed results have been obtained with in vitro systems the application and activity in vivo can with difficultly be overlooked, since many of these polymers are toxic and the estimation of the curing effect besides the toxic effects is hard to accomplish.

As an example it may be pointed out that poly(methacrylic acid) is physiologically active. It has been shown by Ormai and Palkovits (*49*) that under the influence of poly(methacrylic acid) the number of lymphocytes in the blood is increased and the immunization activity of the lymphocytes is intensified.

The utilization of these effects, which have been discovered not only with normal cells but also with malignant tumor cells, seems, however, not to be possible, since the toxicity of poly(methacrylic acid) (*50*) is too high.

A review article on synthetic polymers with antiviral activity and their clinical application has been published by Tarasov *et al.* (*51*).

3.5. Synthetic Substitutes of Plasma

Since substitutes of plasma are not allowed to interact with the human body they are strictly speaking not included in drugs but rather in implantates. Although they ought to be as untoxic as possible, the feasability of their degradation, their metabolism and their excretion are of great interest, because such substances offer themselves as carriers of drugs. Here, the pharmacological behavior of substitutes of plasma will be compared with that of the polymeric drugs, mentioned so far.

The colloid osmotic pressure of plasma substitutes must correspond to that of the plasma and has to stay long enough with the vascular system. Poly(vinylpyrrolidone) (PVP), already described in 1943 by Hecht and Weese (*52*) and synthesized by Reppe (*53*), is still used today besides the half synthetic products dextrane and gelatine. Due to its extensive application during many years the pharmacology of this substance is well studied. Further synthetic polymers have been proposed as plasma substitutes in the following years. Poly(vinyl alcohol) which is easier to prepare than PVP, has been examined in detail. Extensive investigations, particularly in Russian hospitals, have shown the product to be very useful if it is not branched and does not contain parts with a molecular weight $> 40\,000$.

Low molecular parts with a molecular weight $< 15\,000$ are excreted completely by the kidney within a few hours like the PVP preparations are. Since low molecular substances present in the blood are strongly adsorbed at such polymers these polymer fractions are suitable for the depuration of the blood. Thus, the action of snake-poisons and tetanustoxin (*54, 55*), can strongly be reduced if the solution of the polymer is injected a short time after the poisoning.

Recently, further synthetic polymers have been proposed as plasma substitutes. Thus, Kopeček and Bazilova (*56*) examined poly[N-(2-hydroxypropyl)methacrylamide] [*12*].

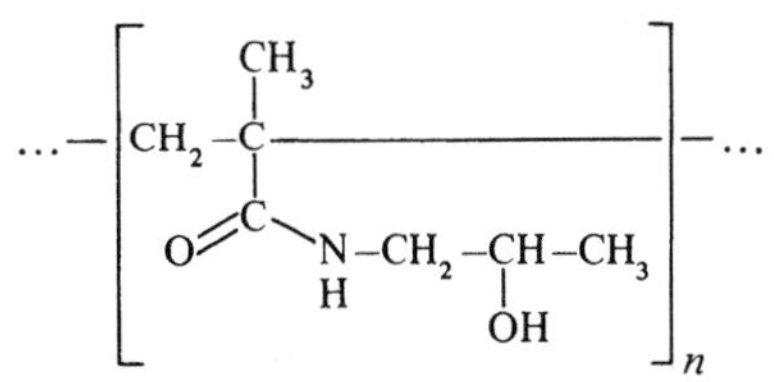

12

This polymer is excreted completely after 5 days (*57*) in the case of molecular weights $< 25\,000$. α, β-Poly [(2-hydroxyethyl)-D, L-aspartamide] has been prepared by Neri and co-workers (*58*) and its usefulness as plasma substitute has been investigated. Polymers *13* with a molecular weight between 10000 and 90000 have been obtained by heating asparagus acid over phosphoric acid at 180 °C and by subsequent reaction with ethanolamine.

Both polymers *12* and *13* are not toxic; the LD_{50} [5] of *13* could not be determined. This substance also shows no action on enzymes and other body specific

---

[5] The LD_{50} (Dosis Letalis 50) indicates at which dosis 50% of the animals die (*59*).

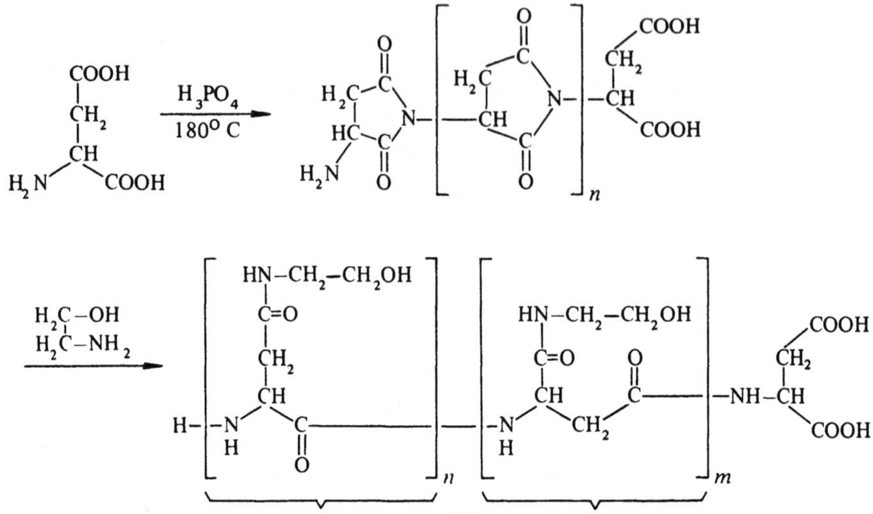

13

A = repeating unit in α-position
B = repeating unit in β-position

substances and also no formation of antigens. After 8h 40% of the originally injected amount of *13* are still in the blood circulation in contrast to gelatine preparations which are degraded and excreted much faster. Histological examinations gave no indications of deposits or alterations in the different organs.

Finally it should be mentioned that copolymers from ethylene oxide and propylene oxide (*60*) with a molecular weight between 50000 and 100000 have been proposed as blood substitutes.

4. Pharmacological Properties of Polymers

Vast pharmacological investigations have so far been carried out only with the preparations PVP and PVNO, which had already found clinical applications. For other polymers there are only scattered data which, however, do not show any particularities in comparison with PVP and PVNO.

The kidney has to be considered as the principal organ for excretion; this holds for the substitutes of plasma as well as for PVNO. Thereby the most important factor for the excretion seems to be the molecular weight. Chemical properties, however, also seem to exercise a certain influence. Thus, *e.g.* up to 50% of a PVP preparation with an average molecular weight[6] of 50000 is excreted (*61*) within

[6] Here, it should be recalled that the average molecular weight indicates only an average value of a more or less broad molecular weight distribution, depending on the method of its determination. For pharmaceutical application great efforts should be made to reach a distribution as narrow as possible, since only in such cases the above statements are conclusive.

24h, whereas in the case of a PVNO preparation (*62*) with a comparable average molecular weight only 1 to 2% are excreted within the same time. A PVP preparation with a molecular weight > 100000, however, is not excreted in measurable quantities.

Although, for that reason, PVP preparations with an average molecular weight between 30000 and 40000 may normally be applied, an increasing amount of higher molecular constituents remain in the body after a considerable period of application. These remainders are deposited (*63*) chiefly in cells of the RHS system, particularly in the liver, in the milt, in the lymphatic gland, and in the lungs and then slowly excreted. According to Alstädter (*64*), alterations disturbing the function of these organs by deposition have not been discovered, whereas Schulze and Rex (*65*) have reported on such.

The distribution of PVNO in the human body has been determined by Grundmann (*66*) by means of ^{14}C-audioradiography. In the case of intravenous application the highest concentration in the blood is reached after 20h; a characteristic distribution in the tissue exists after 3 weeks, which is maintained at least for 1 year. Similarly to the PVP the storage in the RHS-system preferably takes place in the macrophages of the lungs, the liver, and the milt as well as in the submucous intestinal tissue. The polymers are probably absorbed into the cells by phagocytoses and are stored there in the lysosomes or at the membrane of lysosomes.

A similar distribution has been discovered by Ringsdorf and his co-workers (*67*) for polymeric antiradiation substances after the first days of application, also with the help of audioradiography. After 14 and more days, however, a strong cumulation is found in the bone-tissue, most likely reducible to the chemical properties of these sulfur containing polymers.

To sum up it can be said that the above-mentioned different synthetic polymers are stored in the cells of the RHS system with strong phagocytoses activity, however, to a different degree. The different kind of distribution in the organs of the different polymers can so far not be explained in a satisfying way.

It has been shown by several authors that, on intravenous application, in many cases an interaction with the blood substances takes place. Thus, PVNO on intravenous application shows a pharmacological behavior somewhat different from that on subcutaneous application. A partially N-oxidized poly(allyldiethylamine) reacts with heparin with the formation of a complex (*68*).

In all cases the blood-liquour barrier obviously cannot be passed; the brain and the central nervous system remain free. A resorption from the alimentary tract and, therefore, an oral application seem also impossible. A resorption of polymers from the alimentary tract is only possible during the first days after birth, as it has been shown by Malinowski and his co-workers (*69*) with guinea-pigs and rabbits serving as examples.

At the end of this chapter on the pharmacology of polymers it still may be pointed out that investigations of Schmähl (*70*) did not offer any indication on cancerogenous properties of PVNO with mice and rats.

5. Polymeric Drugs Formed by Covalent Bond of Known Active Components to Soluble Macromolecular Carriers

Among the substances which are described here, the polymers have only a vehicle function or they represent the deposit form. Besides the above-described polymeric N-oxides and plasma substitutes further water-soluble nontoxic polymers, such as polymers from 2-hydroxyethyl acrylate or methacrylate (*71*), alkylsulfinyl acrylates and methacrylates (*72*), diethoxyethylenes (*73*), and other substances as yet pharmacologically tested, are in question. Differentiation, however, can be between preparations, whose active substance has to be split off – a fact which has to be taken for granted in the case of substances that are active in the central nervous system – and substances that are still active at the polymeric carrier. Only recently it has been shown by Kaplan (*74, 75*) that this is possible, at least in such cases where an activity is to be supposed via receptors in the cell membrane. In many other cases this question cannot be cleared up unequivocally.

In the following the polymer-bound active compound shall be decisive for the classification and discussion of these types of polymeric drugs.

5.1. Substances which are Active in the Central Nervous System

For preparations like mezcalin, histamine, serotonin, thyroxine, it has been shown by Jatzkewitz (*11, 76*), with mezcalin serving as an example, that in the organism they cannot be split off from the polymeric carrier if they are linked by an amide-like bond to a copolymer from vinylpyrrolidone and acrylic acid, and consequently they cannot be identified anymore either by their activity or by their excretion. By insertion of a so-called spacer group in the form of a dipeptide, as *e.g.* glycine-leucine, the splitting from such a polymer (14) becomes possible again.

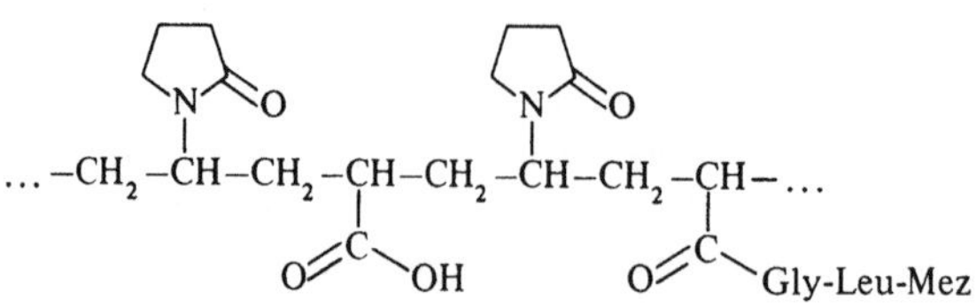

14

In contrast to free mezcalin, which can only be traced during about 20h in the urine, a mezcalin bound in such a manner to a polymeric carrier can still be identified there after 16 days.

Besides substances linked to a polymeric carrier by amide-like bonds, drugs with ester-like bonds cannot be split off. Thus, Kratzl and Kaufmann (*77*) have not detected any splitting or liberation of the nervous system attacking analepticum Ethamivan (Emivan®, Vandid®) in the physiological system, if it is bound to starch through a carboxylic ester bridge which is very easily hydrolyzed in vitro (*15*).

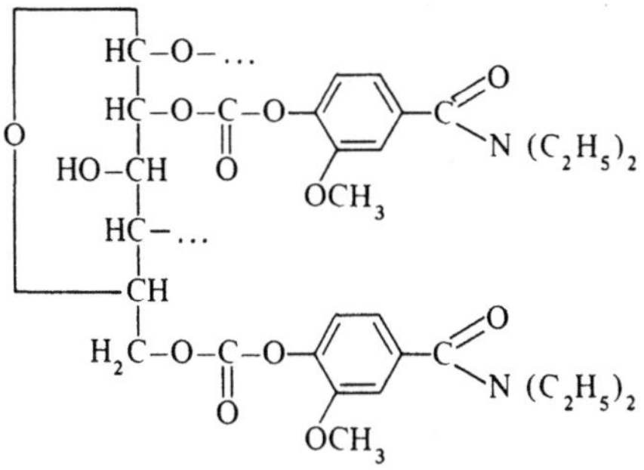

15

It has not been examined if the antiphlogistical activity of various amide-like bound derivatives of aniline and 4-aminoantipyrine [cf. *16*[7]], described by Kotenko (*78*) has to be attributed to their splitting from a poly(methacrylic acid) chain or to secondary effects. Such substances are reported to be extremely nontoxic.

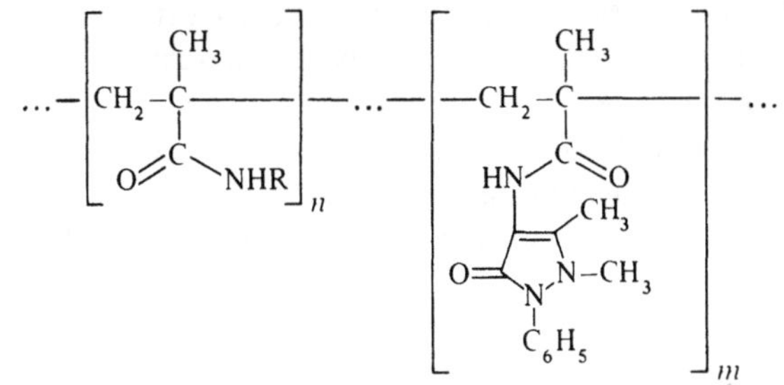

16

A passage of the blood-brain barrier has, however, not to be expected due to their polarity and molecular size. Due to the particular structure of the brain capillaries only lipoid-soluble low molecular substances are able to enter the brain vessels from the blood. This fact is called the blood-brain or blood-liquor barrier which indeed cannot be overcome by polymers but by free low-molecular active substances like mezcalin or aminoantipyrine.

A considerable extension of the duration of activity together with a strongly reduced toxicity has been found (*79*) for a preparation with morphine-antagonistic activity [*17*], the place of its action being localized also in the central nervous system. The possibility of cleavage is in this case attributed to the activated ester, bound between the polymeric chain and the active substance. Besides the above-mentioned derivative of acrylic acid a derivative of vinyl carbamate has also been applied.

[7] Formula *16* and all other formulae of copolymers composed of two or more repeating units are cases of statistical copolymers (and are not blockcopolymers), based on the mode of their preparation. In most of these cases more detailed information is not available.

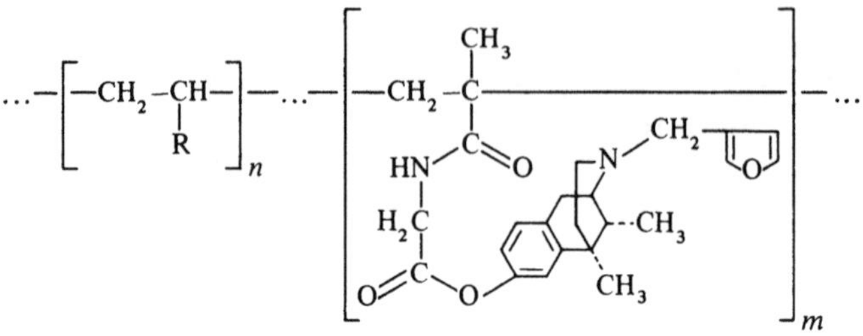

17

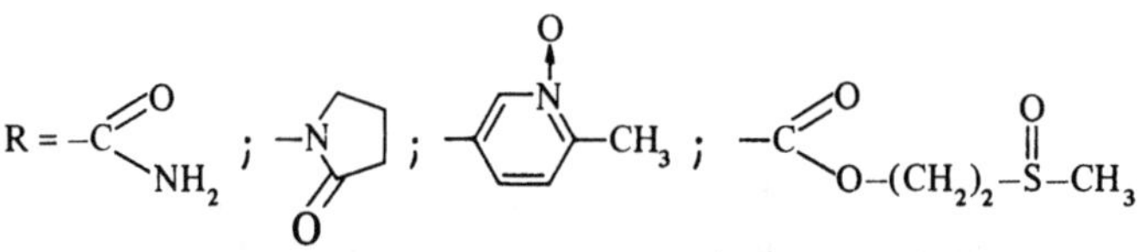

5.2. Antibacterial Agents Bound to Polymeric Carriers

Antibacterial agents linked to polymers have been investigated by various teams. In this case pharmacological examination is easier, since screening tests are often sufficient. However, the results of such *in vitro* tests are not always transferable to *in vivo*-systems and are not always reproducible.

As early as 1962 Ushakov and Panarin (*80*) demonstrated that a derivative of penicillin bound to a copolymer with vinylalcohol and vinylamine (2%) units [*18*] shows an activity which is 30–40 times longer lasting than that of the free penicillin.

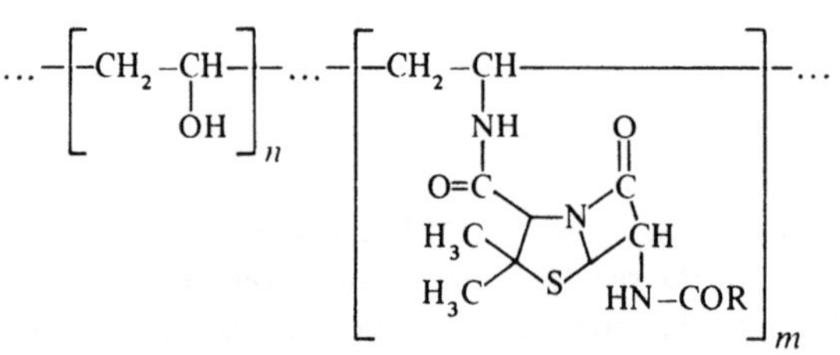

18

Cornell and Donaruma (*81*) have described a polymethacrylate of tropolone [*19*] which, like the free tropolone, is active towards various bacterial strains; the activity of the polymeric preparations being, however, lower than that of the free active substance.

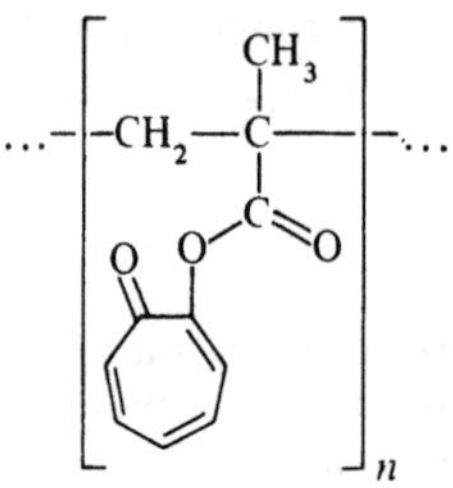

19

Kropachev and his co-workers (*82*) have described a polymeric derivate of sulfanilic amide with a prolonged activity. They prepared copolymers of N-methacryloyl-4-aminobenzenesulfamide with N-vinylpyrrolidone [*20*].

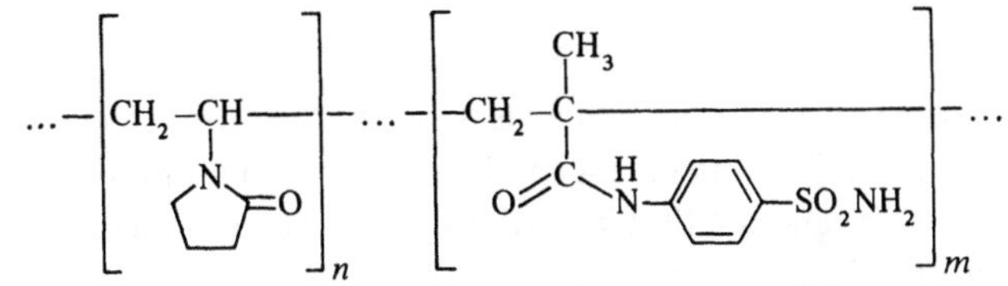

20

On the other hand various polymeric derivatives of 4-aminosalicylic acid show no activity. They are not active either in the form of the poly(N-acryloyl-4-aminosalicylic acid) (*83*) or in the form of the acetylsalicylic ester of starch (*84*). Due to their mechanism of action, that is the interference in the bacterial metabolism, 4-amino salicylic acid and sulfamide derivatives can be active only in the free form.

Kern (*85*) and Scherhag came to this result some years before, when they tested a series of polymers with linked sulfamide groups, which they have synthesized. These two last-mentioned studies seem to confirm the assumption that in the case of a stable covalent bond the cleavage from the polymeric carrier is impossible. In the case where results deviate from this assumption the question arises if parts, which are not covalently bound, had been present as impurities in the polymeric preparations. This may also be the reason why no biological activity is stated for an additional derivative of 4-aminosalicylic acid, a poly(vinylalcohol) [*21*] partially esterified with O-acetyl-4-acetylaminosalicylic acid, described by Ushakov (*86*) in 1964.

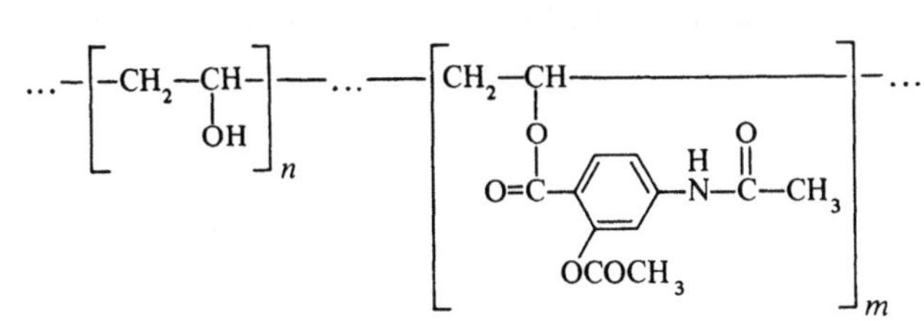

21

In contrast, there are substances like some nitrofurane derivatives for which the presence of particular molecular structures is the decisive condition. Thus, a nitrofurane derivative prepared by Casini and his co-workers (*87*) has shown bacteriostatic properties similar to classical low molecular preparations of nitrofurane, *e.g.* 1-[5-(nitrofurfuryliden)amino]hydantoine. The polymeric substance shows an activity considerably longer than that of the reference substance if parenterally applied, whereas oral application gives no effect. This is easy to understand because, as already mentioned, polymers cannot be resorbed in the digestive tract. Here, the active polymeric substance [*22*] has been prepared by condensation of 5-nitrofuraldehyde with poly(acryloylhydrazide).

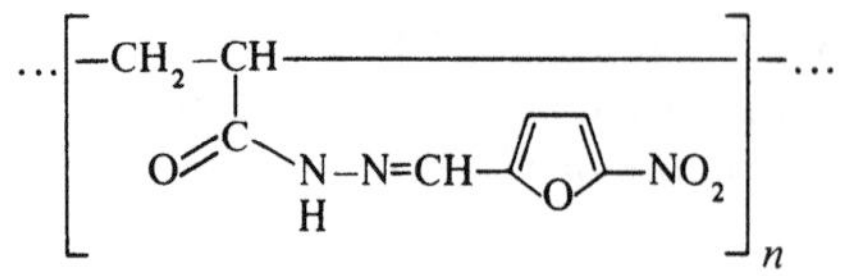

22

The therapeutic importance of such preparations is based on the selective intestinal therapeutics, which above all finds increasing application in the veterinary medicine, since it can be taken for granted that in this case no active substances are able to enter into the body, and thus, into the tissue and flesh.

Recently, additional preparations of this kind have been described; *e.g.* a copolymer from N-allylaminoadamantane and N-vinylpyrrolidone [*23*] (*88*) or poly(acryloyl-9-oxyquinine) and poly(methacryloyl-9-oxyquinine) [*24*] (*89*), respectively.

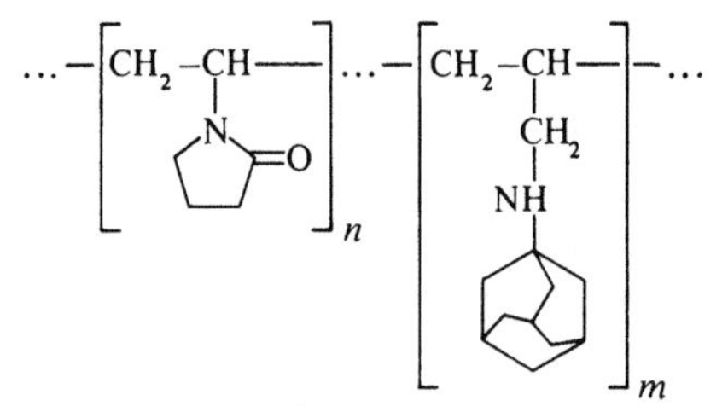

23

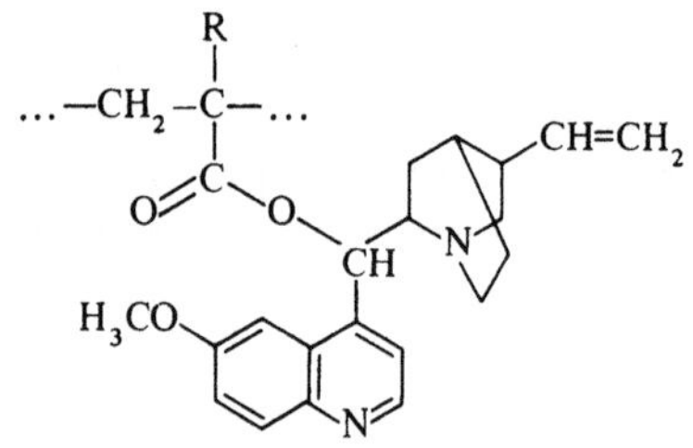

R = H or CH_3 *24*

There still exists a series of further examples of the in vitro-activity of polymeric antibacterials, since such preparations are also applied as protection of polymeric materials against bacterial attack or as pesticides like a poly(methacryloyloxy phenoxarsine) [*25*] (*90*).

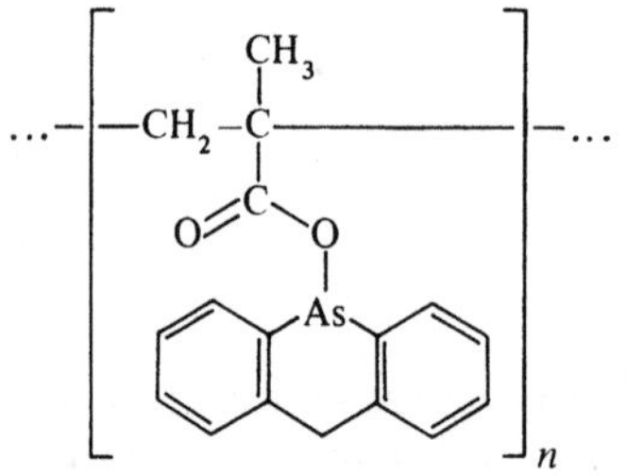

25

5.3. Polymeric Sympathomimetics and Sympatholytic Agents

Sympathomimetic drugs whose effects resemble the responses to stimulation of adrenergic nerves have been transformed and investigated in polymer bound derivatives by different groups.

The syntheses of methacryloyl derivatives of ephedrine [*26*], phenylephedrine [*27*], of salicylic acid and of similar derivatives with methacryloyl chloride have been described by Kotenko and his co-workers (*91*).

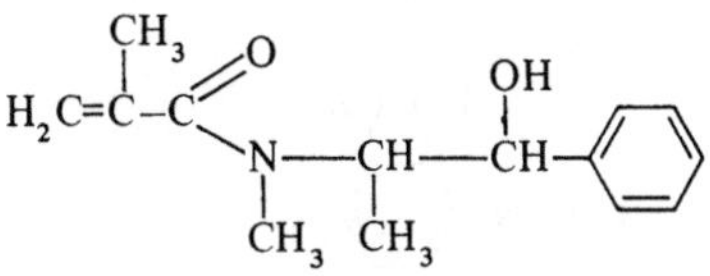

26

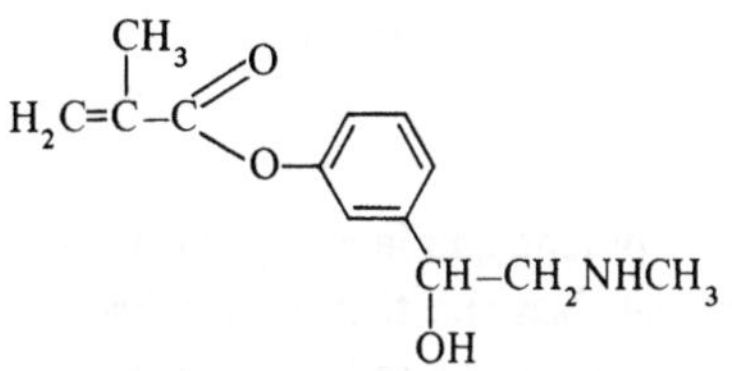

27

These monomers have been copolymerized with N-vinylpyrrolidone. Further polymers have also been obtained by reaction of the "free" drugs with copolymers from maleic anhydride. These polymeric drugs besides an extended activity, show the same or a decreased activity as the low molecular drug.

The monomeric N-methacryloyl derivatives of β-phenylethylamine, amphetamine, ephedrine, and thyramine have been prepared by Zilkha (*92*) and his co-workers in like manner and copolymerized with methacrylic acid, vinyl acetate, and N-vinylpyrrolidone. Besides a distinct prolongation of a factor > 10 chiefly lowered activities have been found. The simultaneously appearing alterations of the activity spectrum can be explained by the supposed hydrolysis in the animal organism only with difficulty.

Particularly, also, for that reason a hydrolytic cleavage of the active substance seems to be most unlikely, because the same group of scientists has been able to show that poly[2-(4-vinylphenyl)ethylamine] [*28*], which certainly cannot hydrolyze, shows also a weak activity. However, this polymer is essentially more toxic than 2-phenylethylamine, and thus, the comparison is difficult (*93*).

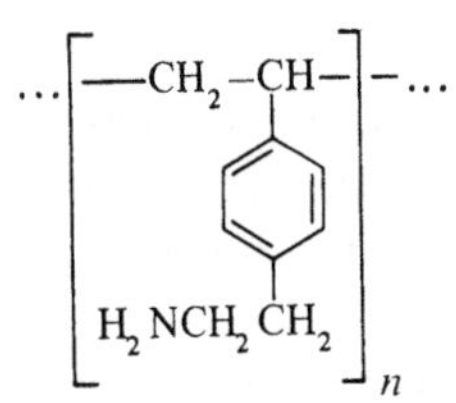

28

The investigations with isoproterenol, noradrenalin, and adrenalin [*29*] covalently bound to beads of glas (20–300 μm of diameter) published recently by Kaplan and his co-workers (*75*), provided also an explication of the results published by Zilkha (*92*).

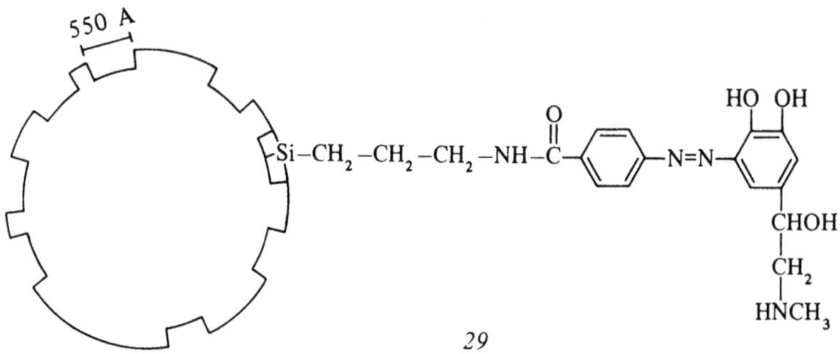

29

In several reaction paths a *p*-diazoniumbenzoylpropylene amide group has been fixed at the surface of the beads prepared from a special glass and then by a subsequent coupling reaction the aromatic ring of the adrenalin was bound to the system. It has been shown that the contact of such covered glass beads with the liberated dog-heart *in vivo* causes a decrease of the heart activity, whereas the isolated myocardium of male cats shows *in vitro* a more intensive and faster contraction. This effect has been eliminated completely only after removal of the glass beads, in which case, however, a slow decrease has already been observed during the contact with the cell membranes. Since it has been shown during the performance of the investigations that no cleavage of the active substance from the glass beads took place, it must be supposed that the effect has to ensue from the action of the receptors of the cell membrane. Analogously this mode of action can also be adopted for drugs linked to polymers.

Since for heart glycosides the same reaction via receptors in the membrane is also supposed, similar studies have been carried out by T. W. Smith and his co-workers (*94*) with digoxin and ouabain bound to a polymer (to albumine or polylysine via different spacers). From the nearly complete disappearance of the activity it had to be concluded these receptors, if localized in the cell membrane, are not present in their surface.

5.4. Hormones Linked to Polymers

Polymeric carriers are of particular importance to peptide hormones, because they are generally degraded very fast in the organism. Accordingly, polymer linked peptide hormones have been proposed and investigated by different groups of scientists [*e.g.* cf. Schuck and Wildi (*95*)].

So far, insulin has been of particular interest; Lehmann and his co-workers (*96*) were able to show recently that insulin linked to a carrier is essentially longer active than the free one.

As here a specific link of the polymer with the terminal amino groups of the peptide is desired, the question of the kind of linkage remains of interest, although there exists a broad practical experience in the domain of carrier linked enzymes and further new methods have been published. Thus, *e.g.* reactive esters, as known from peptide syntheses, offer new possibilities (*97*).

5.5. Polymeric Cancerostatic Agents

In the chemotherapeutic treatment of cancer, polymeric drugs seem to offer some advances in comparison with the hitherto conventional therapeutics. Besides the inhibition of metastases, which has already been discussed (in Chapter 3.2), additional membrane specific interactions should be possible. Due to the differences of the pathological and the healthy cell a cumulation of polymers to which cancerostatic agents are linked could be realized also in the pathological cells, as *e.g.* due to different pH-values or to the higher phagocytosis of the malignant cells. A cleavage of the active substance from the polymeric chain does not seem to be absolutely necessary. Above all, in this range curing effects restrain the objections against toxic effects.

Kargin and his co-workers (*98*) have prepared polymeric derivatives of the cytostatic agent sarcolysin by alkylation of poly(vinylpyridine). Simultaneous alkylation with cetyl bromide yields copolymers with the following structure of chain [*30*].

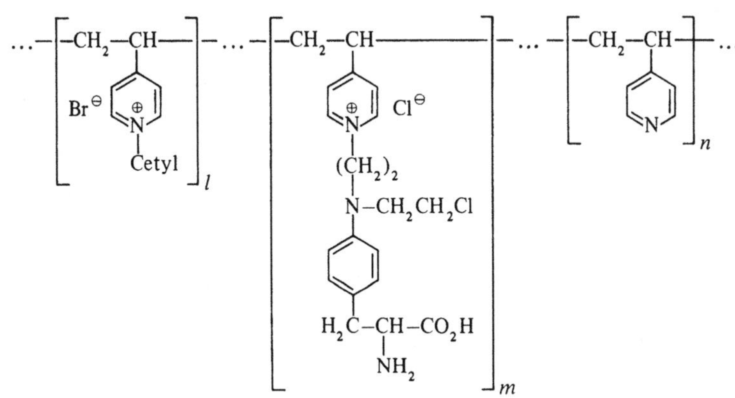

Cetyl. $n-C_{16}H_{33}$ *30*

These polymers are essentially less toxic (factor of 2–10 at LD_{50}) and partially very much longer active than free sarkolysin. An increase in activity, however, is not attained. Here, the mode of action is, however, not clear because bifunctional chloroalkylamine derivatives ("N-Lost" derivatives) seem to be necessary to show some activity. Possibly in this case the alkylpyridinium compound decomposes into a derivative of ethylene or the chloroethyl group is formed again so that a second reactive group is formed besides the chloroethyl residue which is still present.

Analogous to a "N-Lost" derivative, introduced in Germany, the "Endoxane", a polymeric substance [*31*] has also been synthesized (*99*).

So far, no statement about their activity can be made from the pharmacological data.

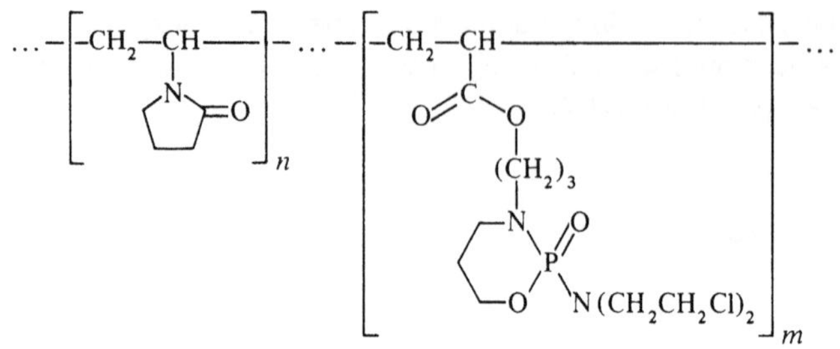

31

Another polymeric cancerostatic agent has also been proposed by Kargin and his co-workers (*100*). In the ethyleneimine derivative of structure *32* the groups acting as alkylating agents are linked to a polymeric chain.

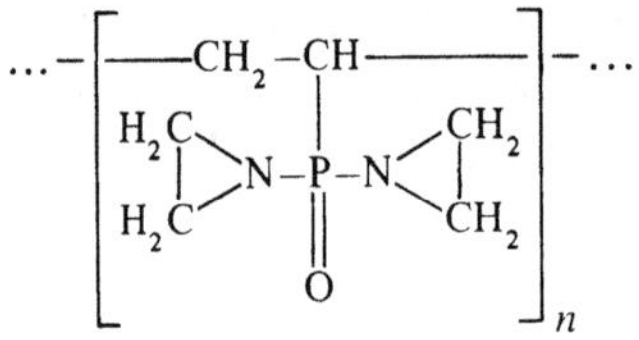

32

A further derivative of ethyleneimine has been described in 1962 by Lidak (*101*). From the monomer [*33*], already active as such, a polymer can be obtained, which is less toxic and has a more effective range.

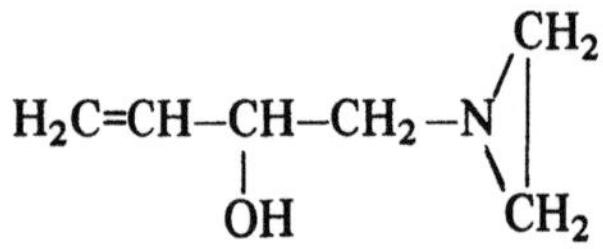

33

Another kind of chemotherapy is based on the assumption that cancer is initiated by radical reactions if cell-specific substances capturing the radicals are lacking. Therefore, as early as 1963 copolymers of vinylpyrrolidone of structure *34* were proposed, containing about 6 mole-percent of vinylhydroquinone units in the chain (*102*).

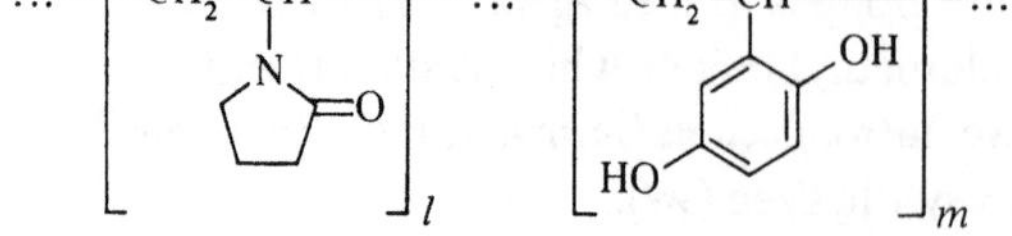

34

Recently, additional radical capturing polymers have been proposed, in which, similar in substances to be discussed later, bifunctional derivatives of the active component are incorporated into polymeric chains directly by polycondensation. Substance *35*, one of three polymers tested by Berlin (*103*), has been found to be the most active one inhibiting most effectively the development of leucosis.

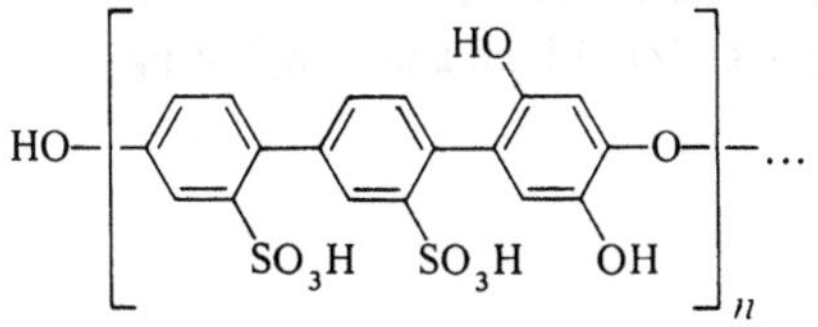

35

Finally, two other polymers may be mentioned in this chapter which are formed by polycondensation of known active substances. A polymeric derivative of fluorouracil with the following formula *36*, synthesized by Schmähl and his co-workers (*104*), showed an improved and extended activity in comparison with the low molecular substance, whereas no activity has been observed by Goodman and his co-workers (*105*) for a peptide-like polymer of sarcolysin.

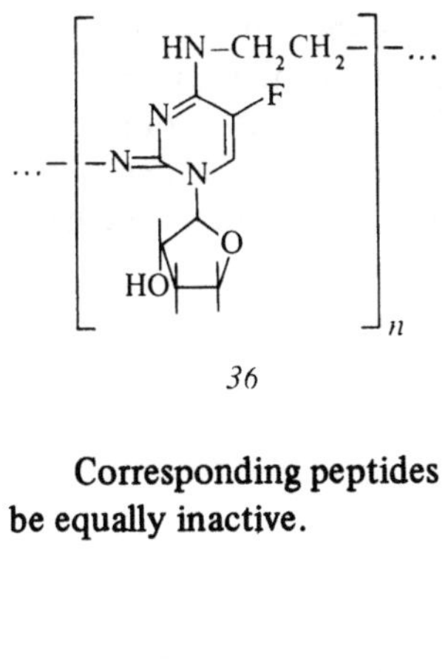

36

Corresponding peptides from sarcolysin and glutamic acid have been found to be equally inactive.

6. Polymers with an Active Substance Incorporated in the Polymeric Chain

As it has already been shown in Chapter 5.5, with *35* as an example, polymeric drugs may also be obtained by direct mutual linkage or via adapters of bi- or oligofunctional molecules of drugs.

The biological activity may be originated by cleavage of such a polymeric chain in the organism, back into the original unit. Then, the time of action should depend on the rate of cleavage. However, since such polymers may possess new physical, biological, and pharmacological properties, which are not comparable either with those of the starting compounds or with those of plasma substitutes or similar preparations, predictions on the intensity and duration of activity of such substances cannot be made. If a polymeric chain is chosen which can be degraded in the organism, even losses of activity have to be taken into account, *e.g.* as it has been demonstrated (*105*) with polypeptide-like sarcolysin.

Baltzly and his co-workers (*106*) have shown already in 1949, that by condensation of formaldehyde with N-[2-(4-methoxyphenyl)ethyl]-N-methylamine in the presence of hydrogen chloride hypotensive agents [*37*] of high and long-lasting activity are obtained.

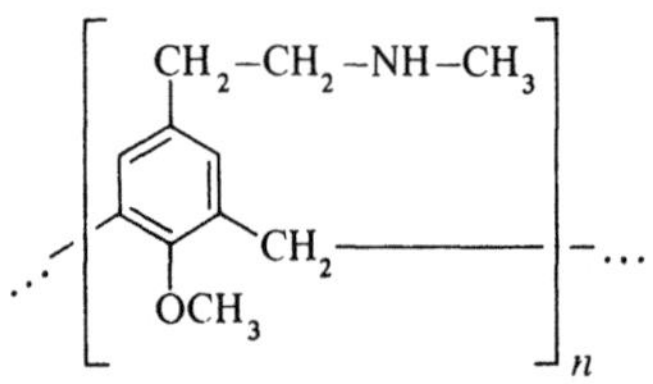

37

These substances in question are oligomers with n = 3–4.

A polymeric preparation has been described, however, by Schueler and Keasling (*107*) in 1956. Starting from the known muscle relaxants which, as it is well-known, have quarternary ammonium groups placed after definite distances in the molecule, they have synthesized polymers in which these quarternary ammonium groups are incorporated in a polymeric chain, as *e.g.* *38*, whereby different numbers of atoms in the repeating unit have been chosen. The toxicity of these polymers increased or decreased independently of the length of the repeating unit. Each of these polymers showed a prolongation of activity with a factor > 10.

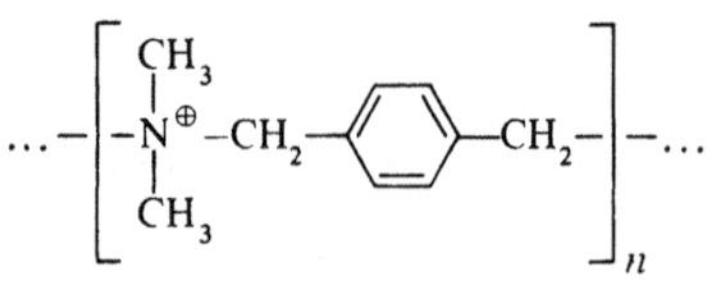

38

The same prolongation of activity has also been found by Fernö and his co-workers (*108*) with a polymeric preparation of estradiol [*39*] by coupling the particular molecule of estradiol via –O–P–O– bridges of phosphoric acid.

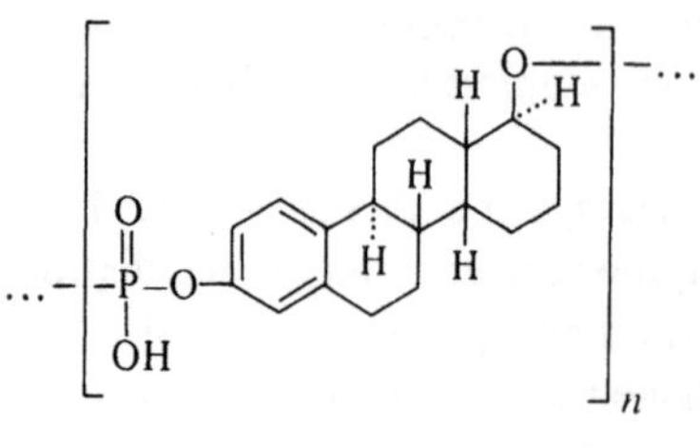

39

Polymers with a molecular weight of about 26000 have been prepared. Their activity is attributed to the slow degradation of the polymeric chain, since after radioactive marking free estradiol could be detected in the urine.

Finally polymeric preparations are of high importance in antiradiation research, since low molecular compounds are often only active for a few minutes. Particularly sulfur groups inside of the molecule seem to be essential for their activity. Ringsdorf and his co-workers (*109*) have recently published an article on various polymeric antiradiation substances which are essentially longer active than the corresponding low molecular compounds. Among others the polymers of the following structure have been tested:

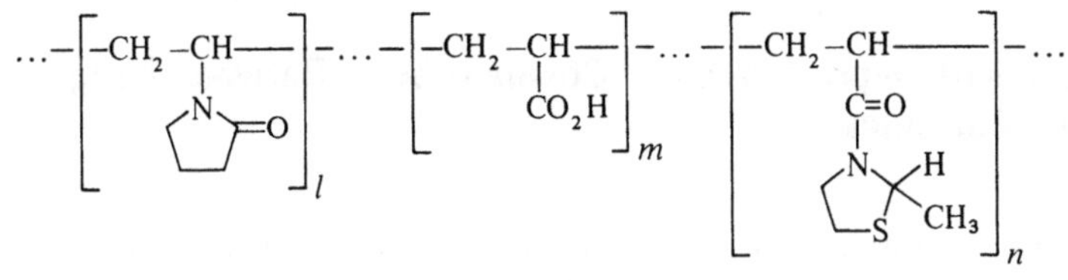

40

The interesting pharmacology and distribution of these polymers in the body has already been pointed out in Chapter 4.

Different types of polymeric sulfamides, *e.g.* showing antimalarial activity, have also been examined in great detail. Donaruma and his co-workers (*110, 111*) have synthesized and tested a considerable number of such substances such as *e.g. 41*.

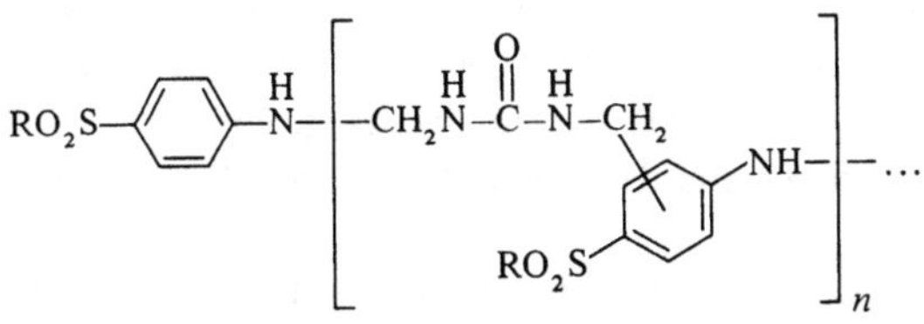

41

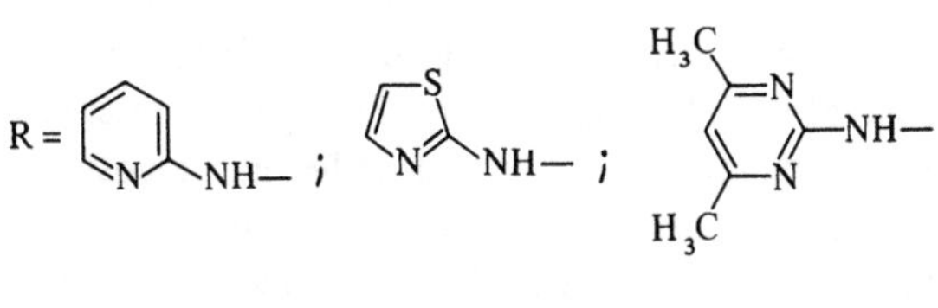

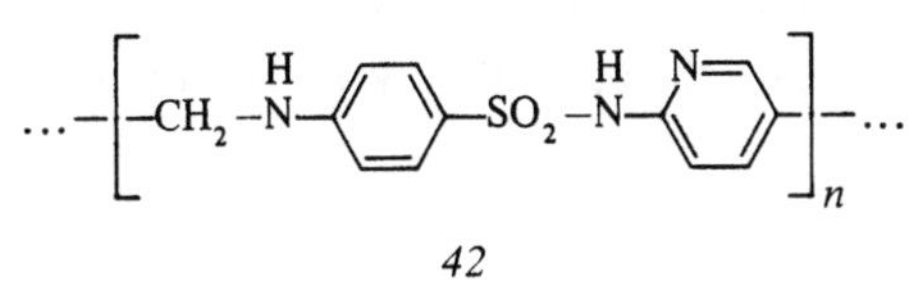

42

In addition, substances like *42* are also bacteriostatically active (*114*). These polymeric sulfamides partially show an extended and higher activity than analogous low molecular preparations.

Polymeric sulfamide derivatives containing a lysine bridge have recently been described by Fenyo and Sclegny (*112*).

Finally it may be pointed out that the preparation of polymeric derivatives of penicillin, by the opening of its lactame ring (*113*) has been tried repeatedly. Since this lactame ring is essential for the activity, the obtained polymers have been inactive or showed only a weak activity against microbes (*114*).

7. Polymeric Forms of Deposit Without Covalent bond Between Drugs and Polymeric Molecules

Besides the polymeric drugs described so far other systems have been and still are intensively investigated – examining the possibility of polymeric deposit forms for low molecular drugs having no covalent bond to the polymer, *e.g.* by admixing of the active substance into the polymer or by imbeding between foils.

The different interactions between both components like polar or van der Waals' forces as well as attainable effects like the extension of activity and altered ability of resorption are discussed by Keipert *et al.* (*115*) in a comprehensive report. Such deposit systems are of great importance for preparations administered per os since the attainable prolongation of action lies in the order of the stay period in the alimentary tract and, therefore, the active substance is set free before the deposit system is excreted.

Although it has already been shown (*12*) with a great number of preparations that only a prolongation of activity by a factor of 2 to 3 can be obtained by such a salt-like bond of the active substances to polymers with ion exchange character, several of such systems have been examined and described (*116–118*), also during the last years.

Preparations whose active substance is imbedded between polymeric foils, have recently obtained some importance in their technical application. Such deposit systems may be inserted into cavities of the body *e.g.* into the uterus or below the palpebra (*119*). The active substance then slowly diffuses through the foil so that its concentration and duration of activity can be regulated by the type and thickness of the membrane (*120*). An essential advantage of this procedure lies in the possibility to remove easily, at any time, the deposit system at the end or to interrupt the treatment. With analogous implantations this is only possible by an operative treatment.

A biologically degradable and resorbable polymeric deposit system has recently been proposed by Yolles (*121*) as an alternative on the basis of polylactide or polyglycolide.

8. Discussion and Outlook

On one hand the progress in biochemistry and molecular biology may enlarge the knowledge on the interactions between natural and synthetic macromolecules but on the other hand the progress in polymer science may, in addition, open new possibilities to synthesize macromolecules as desired. The biological activity of known and new polymeric substances has to be tested.

For the further development of polymeric drugs it will first be necessary to check and guarantee the hitherto existing and partially contradictory results. Thus, in spite of many positive results in the investigation of compatibility or long-lasting activity no decisive breakthrough for a clinical application of polymeric drugs has so far been realized. Preparations like the antisilicosis reagent, poly(vinylpyridine-N-oxide), however, are already in clinical proof for long time, whereas preparations with deposit effect are still in the stage of experiments with animals. Many questions concerning the chemical synthesis, like the reproducibility of composition and molecular weight, and also the pharmacology, like degradation and excretion, have so far not been clarified satisfactorily and need further intensive studies. Just how difficult statements are in this case is demonstrated by two new papers which appeared after the writing of this review. In one case Zilkha and his co-workers (*122*) show, that the transformation of a known drug into a polymer must not necessarily always be connected with a prolongation of activity. The polybarbiturates described by them, indeed retain their pharmacological properties with small changes, but they don't show any extended activity.

In the second paper by Oster and his co-workers (*123*), the hitherto accepted assumption that polymeric substances cannot be resorbed from the alimentary tract is jeopardized. They have been able to detect xanthine-oxidases of beef in the human organism, an enzyme, which is found in the milk and which can have been resorbed only from the alimentary tract. Therefore, it has to be supposed that at least certain proteins can be resorbed from there.

Finally it must be pointed out that in nearly all papers that have hitherto appeared the immunogeneous properties of the polymeric drugs have not been taken into consideration, although for long time it is already known that drugs linked to a carrier are able to act as antigenes (*124*).

For the induction of antibody formation and their subsequent isolation a great number of substances like the peptide hormone insulin and angiotensin (*125*) or the steroides 17-β-estradiol (*126*) and digitoxin (*127*) has been coupled to natural polymers, *e.g.* to serum albumin of beef, but also to synthetic polymers like polylysine. Since, however, such an immunogenity of polymeric drugs is highly undesired and no antibody formation should occur, because antibodies abolish the pharmacological activity of a substance by complex formation, it seems to be necessary, to set emphasize of the future research here. The question remains if it will be possible to lower or to abolish such immunogeneous properties of polymeric drugs by alteration of the polymeric carrier, of its structure, its molecular weight, or its degradability. Here, a vast field for new studies is opened.

9. References

1. a) Schuller, H.: lecture given at the "Tagung der Naturforscher und Ärzte", München 1972.L. b) Donaruma, L. G. Progress in Polymer Science **4**, 18 (1974).
2. Lymann, D. J.: Angew. Chem. **86**, 145 (1974).
3. Mark, H. F.: Pure Appl. Chem. **16**, 201 (1968).
4. Lymann, D. J.: Rev. Macromol. Chem. **1**, 355 (1966).
5. Kovanov, V. V.: Usp. Khim. Tekhnol. Polim. **1970**, 176.
6. Yabbločkin, V. D.: Korm. Biol. Med. **3**, 41 (1969).
7. Dyck, M. F., Winters, P. R.: J. Biomed. Mater. Res. **5**, 207 (1971).
8. Schuerch, C.: Advan. Polym. Sci. **10**, 173 (1972).
9. Schlipköter, H. W., Brockhaus, A.: Deut. Med. Wochenschr. **93**, 2479 (1968).
10. Merigan, T. C.: Nature (London) **214**, 416 (1967).
11. Jatzkewitz, H.: Hoppe-Seyler's Z. Physiol. Chem. **297**, 149 (1954).
12. Khomyakov, K. P., Virnik, A. D., Rogovin, Z. A.: Russ. Chem. Rev. **33**, 462 (1964).
13. Tadashi Nakata. Kobunshi: (High Polym. Japan) **18**, 98 (1969).
14. Kotenko, S. I., Linsunkin, V. I.: Farm. J. (Kiev) **26** (1), 17 (1971).
15. Kabaiwanow, W., Georgiewa, M.: Plaste Kautschuk **19**, 886 and 904 (1972).
16. Danusso, F., Ferruti, P.: Chim. Ind. (Milano) **55**, 147 (1973).
17. Ushakov, S. N.: "Synthetic Polymers Applied in the Medicine", State publishing house of medicine, Moscow 1961.
18. Schlipköter, H. W., Brockhaus, A.: Deut. Med. Wochenschr. **85**, 920 and 933 (1960).
19. Schlipköter, H. W., Brockhaus, A.: Klin. Wochenschr. **39**, 1182 (1961).
20. Dolgner, R., Brockhaus, A., Schlipköter, H. W.: Beitr. Silikose-Forsch., Sonderbd. Grundfragen Silikoseforsch. **6**, 213 (1965).
21. Klosterkötter, W., Gono, F.: "Ergebn. Untersuch. a. d. Gebiet d. Staub- u. Silikosebekämpfung im Steinkohlebergbau", Bösmann, Detmold 1969, Vol. 7, p. 159.
22. Antweiler, H., Djie, T. T.: Beitr. Silikose-Forsch. **23**, 61–90 (1971).
23. Ferruti, P., Marchisio, M. A.: Medicina del Lavoro **57**, 481–491 (1966).
24. Holt, P. F., Lindsay, H., Beck, E. G.: Brit. J. Pharmacol. **38**, 192–201 (1970).
25. Schlipköter, H. W., Brockhaus, A.: Fortschr. Staublungenforsch. **1963**, 397.
26. Antweiler, H.: Beitr. Silikose-Forsch. Sonderbd. Grundfragen Silikose-Forsch. **5**, 153–169 (1963).
27. Manojlovic, "Ergebn. Unters. a. d. Gebiet d. Staub- und Silikosebekämpfung im Steinkohlebergbau", Bösmann, Detmold 1969, Vol. 7, p. 179–183.
28. Holt, P. F.: Brit. J. Ind. Med. **28**, 72 (1971).
29. Ferruti, P., Danusso, F., Franchi, G., Polentarutti, N., Garatini, S.: J. Med. Chem. **16**, 496 (1973).
30. Franchi, G., Morsca, L.: Reyers-Degli-Innocenti, I., Garatini, S.: Eur. J. Cancer 7, 533 (1971).
31. Calvin, M.: Angew. Chem. **86**, 111 (1974).
32. Regelson, W.: #Chem. Eng. News 47 (97), 35 (1969).
33. Cornforth, J. W., D'Arcy Hart, P., Stock, J. A.: Nature (London) **168**, 150 (1951).
34. Cornforth, J. W., Morgan, E. D., Potts, K. T.: Tetrahedron **29**, 1659 (1973).
35. Dittmann, E. Ch.: Arch. Pharmacol. **276**, 199 (1973).
36. Weiner, B. Z., Zilkha, A.: J. Med. Chem. **16**, 573 (1973).
37. Gabor, S., Zugravu, E., Anca, Z., Papilian, V. V.: Igiena **21**, 329 (1972).
38. Regelson, W.: Advan. Exp. Med. Biol. **1**, 315 (1967).
39. Breslow, D. S., Edwards, E. I., Newburg, N. R.: Nature (London) **246**, 160 (1973); Breslow, D.: Intern. Symposium on Macromolecules, Jerusalem 1975, Abstracts, p. 9.
40. Field, A. K., Tytell, A. A., Lampson, G. P., Hilleman, M. R.: Proc. Nat. Acad. Sci. USA **58**, 1004 (1967).
41. Vileck, J., Barmak, S. L., Havell, E. A.: J. Virol. **10**, 614 (1972).
42. Brit. Pat. 051514, DOS 2 253 977 v.10. 5. 1973; inventor: Harnden M. R.
43. Pitha, P. M., Teich, N. M., Lowy, D. R., Pitha, J.: Proc. Nat. Acad. Sci. USA **70**, 1204 (1973).

44. Reynolds, F., Grunberger, D., Pitha, J., Pitha, P. M.: Biochemistry **11**, 3261 (1972).
45. Somer, P. De: #Int. Symp. Standardization of Interferon and Interferon-Inducers, London 1969, Symp. Ser. Immunbiol. Standard. **14**, 221, Karger, Basel/New York 1970.
46. Billiau, A., Mugembe, J. J., Somer, P. De: Nature (London) New Biol. **232**, 183 (1971).
47. Abeels, M. J. F.: Experientia **26**, 483 (1970).
48. Tempel, K.: Arzneim.-Forsch. **19**, 1611 (1969).
49. Ormai, S., Palkovits, M.: Blut **24**, 161 (1972).
50. Ormai, S., Hagenbeck, A., Bekkum, D. W. van: Eur. J. Cancer **6**, 365 (1970).
51. Tarazov, V. N., Bresler, S. E., Kropachev, V. A., Trukhmanova, L. B., Vilner, L. M., Zeitlenok, N. A.: Tr. Inst. Poliomielita Virus Entsefalitov Akad. Med. Nauk SSSR **16**, 308 (1971).
52. Hecht, G., Weese, H.: Münch. Med. Wochenschr. **85**, 11 (1943).
53. Reppe, W., Hecht, G., Weese, H.: DRP, Nr. 738 994.
54. Beuchelt, H.: Therapeut. Ber. (Bayer) **10**, 225 (1950).
55. Ansel, R., Intorp. H., Weisschedel, E.: Arzneim.-Forsch. **13**, 949 (1963).
56. Kopeček, J., Bažilova, H.: Eur. Polym. J. **9**, 7 (1973).
57. Kopeček, J.: personal communication.
58. Neri, P., Antoni, G., Benvenuti, F., Cocola, F., Gazzei, G.: J. Med. Chem. **16**, 893 (1973).
59. Mutschler, E.: "Arzneimittelwirkungen". Stuttgart: Wissenschaftl. Verlagsges., 1970, p. 95.
60. S.-African Pat. 6 805, 978 (1969), Wyandotte Chemicals Corp.; inventor: Hymes, A. C.; C. A. **72**, 39088 (1970).
61. Hecht, G.: Therapeut. Ber. (Bayer) **10**, 264 (1956).
62. Liefländer, M., Strecker, F. J.: Hoppe-Seyler's Z. Physiol. Chem. **347**, 268 (1966).
63. Gresser, A.: Deut. Med. Wochenschr. **88**, 2217 (1963).
64. Alstädter, R.: Ärztliche Sammelblätter **49** (1) (1960).
65. Schulze, E., Rex, H.: Z. ärztl. Fortbild. **62**, 745 (1968).
66. Grundmann, E.: Fortschr. Staublungenforsch. **2**, 223 (1967).
67. Ringsdorf, H.: lecture given at the 13th Microsymposium on Macromolecules in Prag, 1973.
68. Marchisio, M. A., Sbertoli, C., Farina, G., Ferruti, P.: Eur. J. Pharamcol. **12**, 236 (1970).
69. Malinowska, K. W., Hardy, R. N., Nathanielsz, P. W.: J. Endocrinol. **55**, 397 (1972).
70. Schmähl, D.: Arzneim.-Forsch. **19**, 1313 (1969).
71. Šprincl, L., Vacik, J., Kopeček, J.: J. Biomed. Mater, Res. **7**, 123 (1973).
72. Batz, H. G., Hofmann, V., Ringsdorf, H.: Makromol. Chem. **169**, 323 (1973).
73. Ringsdorf, H., Batz, H. G., Daniel, H., Merz, H.: unpublished results.
74. Venter, J. C., Ross, J., Dixon, J. E., Mayer, S. E., Kaplan, N. O.: Proc. Nat. Acad. Sci. USA **70**, 1214 (1973).
75. Venter, J. C., Dixon, J. E., Maroko, P. R., Kaplan, N. O.: Proc. Nat. Acad. Sci, USA **69**, 1141 (1972).
76. Jatzkewitz, H.: DAS 1 041 052 vom 7. 7. 1954; Z. Naturforsch. **10b**, 27 (1955).
77. Kratzl, H., Bertl, H., Kaufmann, E.: Monatsh. Chem. **92**, 384 (1961).
78. Kotenko, S. I.: Khim. Farm. Zh. 7, 27 (1969); UdSSR Pat. 296778 (1971); inventors: Kotenko, S. I., Fadeichera, A. G., Pochink, V. Ya., Mokhort, N. A.; C. A. **75**, 89062 (1971).
79. Batz, H. G., Daniel, H., Franzmann, G., Koldehoff, J., Merz, H., Ringsdorf, H., Stockhaus, K.: in preparation.
80. Ushakov, S. N., Panarin, E. F.: Dokl. Akad. Nauk SSSR **147**, 1102–1104 (1962); **149**, 334–337 (1963); C. A. **58**, 11168 (1963); **59**, 6201 (1963).
81. Cornell, R. J., Donaruma, L. G.: J. Polym. Sci. Part A, **3**, 827 (1965); J. Med. Chem. **8**, 388 (1965).
82. UdSSR Pat. 328110 (1970); inventors: Shchukovskaya, L., Kapustayanskaya, A. M., Kropachev, V. A.; C. A. 77, 20349 (1972); UdSSR Pat. 332 103 (1970); inventors: Shchukovskaya, L., Kaputtayanskaya, A. M., Kropachev, V. A; C. A. **77**, 225 1 27 (1972).
83. Kennedy, J. F., Epton, J., Kennedy, G. R., Antimicrob. Ag. Chemother. **3**, 29 (1973).
84. Kratzl, K., Kaufmann, E., Monatsh. Chem. **92**, 372 (1961); Kratzl, K., Kaufmann, E., Kraupp, O., Stormann, H., Monatsh. Chem. **92**, 379 (1961).
85. Kern, W., Schulz, R. C.: Angew. Chem. **69**, 153 (1957).
86. Trukhmanova, L. B., Ushakov, S. N., Markelova, T. M.: Vysokomol. Soedin. **6**, 1346 (1964).

87. Ascoli, F., Casini, G., Ferappi, M., Tubaro, S.: J. Med. Chem. **10**, 97 (1967).
88. Holt, P., Thadani, C.: Makromol. Chem. **169**, 55 (1973).
89. Yamauchi, K. Kinoshita, M., Imoto, M.: J. Macromol. Sci. Chem. **A 6**, 73 (1972); Bull. Chem. Soc. Japan **44**, 3186 (1971).
90. US Pat. 3 660 353 (1972); Dow Chemical Co.; inventors: Wang, C. S., Sheetz, D. P.; C. A. **77**, 75732 (1972).
91. Kotenko, S. I., Linsunkin, Ya. I., Pochinok, V. Ya.: Farm. Zh. (Kiev) **27**, (1), 19 (1972).
92. Weiner, B. Z., Tahan, M., Zilkha, A.: J. Med. Chem. **15**, 410 (1971).
93. Tahan, M., Lender, N., Zilkha, A.: Israel J. Chem. **10**, 835 (1972).
94. Smith, T. W., Wagner, H. Jr., Markis, J. E., Young, M.: J. Clin. Investigation **51**, 1777 (1972).
95. US Pat. 3 679 653 (1972); inventors: Schuk, J. M., Wildi, B. S.; C. A. **77**, 105588 (1972).
96. Krämer, D. M., Lehmann, K.: FEBS Lett. **22**, 49 (1972).
97. Batz, H. G., Franzmann, G., Ringsdorf, H.: Makromol. Chem. **172**, 27 (1973).
98. Zubova, O. V., Kirsh, Yu. E., Lebedeva, T. S., Shorokhova, A. A., Silaev, A. B., Kabanov, V. A., Kargin, V. A.: Dokl. Akad. Nauk SSSR **186**, 477 (1969).
99. Batz, H. G., Ringsdorf, H., Ritter, H.: Makromol. Chem. **175**, 2229 (1974).
100. USSR Pat. 260 887 (1970); inventors: Nadzhimutdinov, Sh., Kargin, V. A., Usmanov, Kh. U., Bruevich, G. Yu.: C. A. **72**, 133 (1970).
101. Lidak, M. J., Gillew, S. A.: "Syntheses and Investigations of Cancerostatic Preparations" State publishers house of medicine. Moscow: 1962, p. 150.
102. Uschakov, S., Klimova, O. M., Kartschmurtschik, O. S., Gmutskaja, E. M.: Ber. Akad. Wiss. UdSSR **143**, 231–234 (1962); C. Z. **134**, 19303 (1963).
103. Berlin, A. A.: Izv. Akad. Nauk SSSR., Ser. Biol. **2**, 294 (1971).
104. Ballweg, H., Schmähl, D., Wedelstaedt, E. von: Arzneim.-Forsch. **19**, 1296 (1969).
105. Mosher, C. W., Iwamoto, R. H., Acton, E. M., Goodman, L.: J. Med. Chem. **7**, 650 (1964).
106. Baltzly, R., Buck, J. S., Beer, E. J. de, Webb, F. J.: J. Amer. Chem. Soc. **71**, 1301 (1949).
107. Schueler, F. W., Keasling, H. H.: J. Amer. Pharm. Ass. **45**, 792 (1956).
108. Fernö, O., Fex, H., Högberg, B., Linderot, T., Veige, S.: Acta Chem. Scand. **12**, 1675 (1958).
109. Ringsdorf, H.: Strahlentherapie **132**, 627 (1967); Ringsdorf, H., Heisler, A. G., Müller, F. H., Graul, E. H., Rüther, W., In: Biological Aspects of Radiation Protection. Arg. T. Sugahora O. Hug. Tokio: Igaku Shoin Ltd. 1971, p. 138.
110. Dombroski, J. R., Donaruma, L. G.: J. Appl. Polym. Sci. **15**, 1219–1230 (1971).
111. Donaruma, L. G., Razzano, J.: J. Med. Chem. **9**, 258 (1966).
112. Baumais, J., Fenyo, J. C., Selegny, E.: Eur. Polym. J. **9**, 15 (1973); Fenyo, J. C.: Eur. Polym. J. **10**, 233 (1974).
113. Smith, H., Marshall, A. C.: Nature (London) **232**, 45 (1971).
114. Butcher, B. T., Stanfield, M. K., Stewart, G. T., Zemelman, R.: Molecular Crystals and Liquid Crystals **12**, 321 (1971).
115. Keipert, S., Becher, J., Schultze, H. H., Voigt, R.: Pharmazie **28**, 145 (1973).
116. USSR Pat. 375 103 (1969); DOS 2 043 330; inventors: Kaschkina, N. A., Pormale, M., Kalnins, A., Susters, J., Dambite, G., Reinberg, J. M., Valkova, I. V., Tipainis, V., Skutelis, A., Jankowskis, S. K., Olinja, I. P.: C. A. **75**, 65 436 (1971).
117. USSR Pat. 317 765 (1969); DOS 2 013 212; inventors: Kaschkina, N. A., Pormale, M., Lanberte, H., Kalnins, A., Susters, J., Mikazans, V., Purvins, I., Skutelis, A.: C. A. **74**, 57 324 (1971).
118. Fr. Demande 2 116 453 (1972); inventors: Kaschkina, N. A., Pormale, M., Lanberte, H., Kalnins, A., Surna, J., Susters, J., Mikazans, V., Purvins, I., Skutelis, A., Miltins, A.: C. A. **78**, 10 2021 (1973).
119. US Pat. 179 129, Appl. 9. 9. 1971; Alza Corp.
120. Leaders, F. E., Hecht, G., Hoose, M. von, Kellog, M.: Ann. Ophtal. **5**, 513 (1973).
121. Woodland, J. H. R., Yolles, S., Blake, D. A., Helrich, M., Meyer, F. J.: J. Med. Chem. **16**, 897 (1973).
122. Tahan, M., Lander, N., Zilkha, A.: Israel J. Chem. **11**, 835 (1973).
123. Oster, K. A., Ostera, J. B., Ross, D. J.: International Laboratory **1974** (5), 15; Ross, P. J., Ptaszynski, M., Oster, K. A.: Proc. Soc. Exp. Biol. Med. 144, 523 (1973).

124. Erlanger, B. F.: Pharmacolog. Rev. **25**, 271 (1973).
125. Vallotton, M. B.: Immunochemistry 7, 529 (1970).
126. Liebermann, S., Erlanger, B. F., Beiser, S. M., Agate, F. J.: Recent Prog. Hormone Res. **15**, 165 (1959); Zimmering, P. E., Liebermann, S., Erlanger, B. F.: Biochemistry **6**, 154 (1967).
127. Butler, V. P., Chen, J. P.: Proc. Nat. Acad. Sci. USA **57**, 71 (1960); Oliver, G. C., Parker, B. M., Brasfield, D. L., Parker, C. W.: J. Clin. Invest. **47**, 1035 (1968).

Received April 29, 1976

Cooperative Actions in the Nucleophile-Containing Polymers

Takeo Shimidzu

Department of Hydrocarbon Chemistry, Faculty of Engineering, Kyoto University, Kyoto 606, Japan

Table of Contents

1. Introduction

Imidazole is found in many enzymes as histidine residue. It plays a role of electron donor in biological systems. Imidazole is one of the nucleophiles, and many nucleophile-containing polymers have been synthesized as a mode of the enzyme such as α-chymotrypsin.

Generally speaking, the role of the enzyme consists of the selective and specific attraction of substrate and the highly efficient catalysis. Every enzyme has its own characteristic feature; for example, the specificity in the binding and a charge-relay action in the catalysis in α-chymotrypsin, the contribution of the imidazole moiety as an electron donor to the electrophilicity of zinc ion in carboxypeptidase, the change in the spin state and the reactivity of the transition metal ion by the coordination of the imidazole in the hemochrome. These typical characteristic features are the result of the cooperative actions of the constituents.

To date, to materialize these characteristic features in the concrete or the abstract, many nucleophile-containing polymers have been synthesized and studied. In particular, the abstract materialization is as innumerable as the functional polymers, taking account of polymers other than the nucleophile-containing polymers. The polymers resulting from the abstract materialization have been occupying their particular fields.

This article describes the cooperative action in which the nucleophile-containing polymer is studied as the model of the enzyme.

In general the catalytic activity of the polymer molecule is inclined to be higher than that of the corresponding low molecular weight compound. It can be assumed that all properties of the polymer influence the nature of the polymer catalyst with a cooperation. The cooperation is classified briefly and considered to be as follows.

In Section 2 mechanisms of a few enzymes are briefly reviewed as a model. α-Chymotrypsin is one of the well-studied enzymes which has the nature of the nucleophilicity in the catalysis. Carboxypeptidase has the nature of electrophilicity in the catalytic center at which the zinc ion is supported and controlled by the imidazole moiety. Hemochrome has the characteristic of the coordination of imidazole with transition metal ion.

In Section 3 cooperative function of non-metalloenzyme model polymer will be discussed from several points of view. The cooperative function can be observed not only in the catalyst for hydrolysis but also in the catalyst for the synthesis.

In Section 4 the nucleophile-containing polymer which is able to chelate with the metal ion will be discussed. In this section the nature of the polymeric substance plays a role in the functionality.

2. The Cooperative Action in Typical Enzymes

2.1. α-Chymotrypsin

Though α-chymotrypsin has not an extremely high selectivity in binding and catalytic activity compared with other enzymes, it is one of the most well-studied enzymes.

The structural analysis has been carried out right up to the recognition of molecular level (*1, 2*). α-Chymotrypsin is poly(amino acid) consisting of 245 amino acids, having relatively deep grooves. It catalyzes the hydrolysis of carboxylic acid derivatives such as protein, simple amides, esters, etc. The active site is composed of aspartic acid, Asp (102) . . . histidine, His (57) . . . serine, Ser (195), and the distances between Asp . . . His and His . . . Ser are 2.8 Å and 3.0 Å, respectively. Electronic structures of these moieties depend on the pH of the reaction system. In the range of pH > 7 at which α-chymotrypsin is active, $-COO^-$ of Asp attracts N_4 proton in imidazolyl of His, and N_1 in the imidazolyl of His attracts the proton in OH of Ser. It is called "charge-relay system".

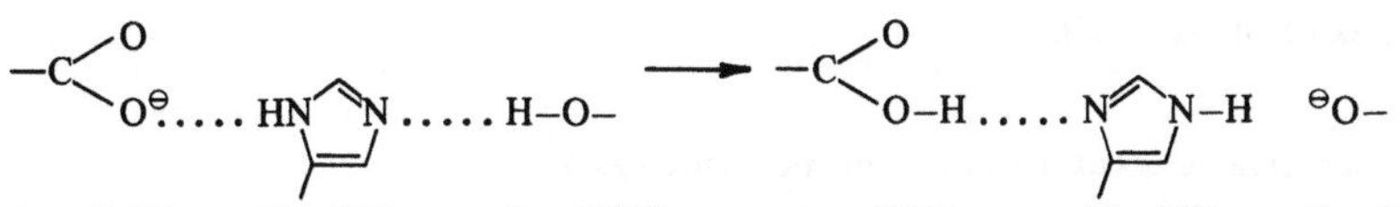

Asp (102) His (57) Ser (195) Asp (102) His (57) Ser (195)

The $-O^-$anion in Ser plays the role of the catalysis so that α-chymotrypsin is a member of the class of enzymes called "serine proteinase". We have many serine proteinase such as tripsin, elastase, subtilicin, etc. Papain is called thiol proteinase and contains thiol moiety in place of serine.

Imidazole moiety in His plays an important role in enzymatic activity in these enzymes as an electron donor. Besides, in the case of ribonuclease A, the imidazole moiety participates directly in catalysis, and in the case of lysozyme, –COOH moiety participates in lieu of the imidazole and cooperates with $-COO^-$.

Reaction course of the ester hydrolysis by α-chymotrypsin is written as follows,

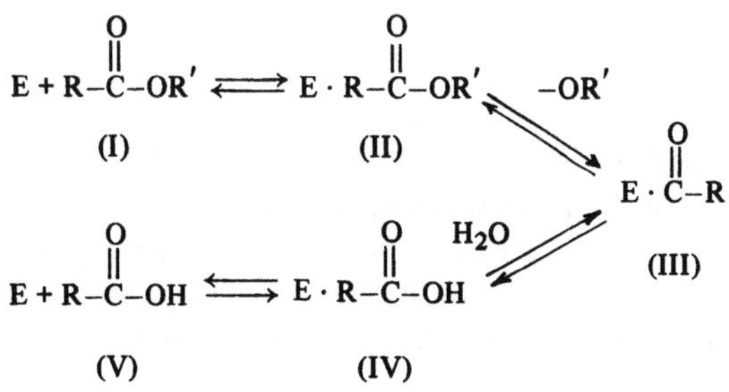

(E: α-chymotrypsin)

The course consists of the acyl process (–OR′ process) and deacyl process, and (II) is called Michaelis complex. The formation of the Michaelis complex which is the predominant process of substrate selection serves the catalysis with high efficiency. Although it depends on pH a drastic decrease in activation entropy in this process shows that the free energy barrier of the Michaelis complex formation is entropic (*3*). The magnitude of the decrease is so large that it can not be explained by a decrease of the freedom due to a steric fit of the enzyme and the substrate, therefore changes in conformation of the enzyme and in structure of water molecules at the binding should be taken into account (*4*).

In addition, as stated above, the catalytic active center forms the charge-relay system and creates a high activity in the catalysis. Recent NMR studies support the concept of the charge-relay system (*5, 6*). By the theoretical calculation the importance of the charge-relay is also discussed (*7*). In both cases where Asp is absent, and where the water molecule is in place of His, the potential barrier of the reaction becomes higher and the substrate plays a role of reduction of the potential barrier of the reaction by 0.21 eV.

Essentially, α-chymotrypsin has these characteristics; the selectivity in the substrate binding, the charge-relay system in the active center and a contribution of the bound substrate to the catalysis, as cooperativities.

In Tabel 1 several reaction parameters in hydrolyses by α-chymotrypsin are shown as an example (*8–10*).

Table 1. Reaction parameters of hydrolyses by α-chymotrypsin

Substrate	K_m (M)	K (M^{-1})	k_{cat} (sec^{-1})	k_{cat}/K_m (M^{-1}/sec^{-1})
N-Acetyl-L-tyrosine ethyl ester	7×10^{-4}	$\sim 1.43 \times 10^{3}$	193	2.7×10^{5}
N-Acetyl-L-triptophane ethyl ester	9×10^{-5}	$\sim 1.11 \times 10^{4}$	50.6	5.6×10^{5}
N-Acetyl-L-phenylalanine ethyl ester	1.1×10^{-3}	$\sim 9.1 \times 10^{2}$	173	1.6×10^{5}
N-Acetyl-glycine ethyl ester	1.0×10^{-2}	$\sim 1.0 \times 10^{2}$	0.008	8.0×10^{-1}
p-Nitrophenyl acetate	4×10^{-5}	$\sim 2.5 \times 10^{4}$	0.012	3.0×10^{2}

K_m, Michaelis constant ($\cong K^{-1}$) : k_{cat}, the first order rate constant (see below).

2.2. Carboxypeptidase

Electrophilic action in enzymes is carried out by metal ion or coenzyme. The metal ion which is a component of the active center is bound to imidazole moiety in histidine by chelation. The typical enzymes are carboxypeptidase and decarboxylase, etc. A change in electronic structure of the metal ion by the chelation plays an important role in the catalysis. The catalytic activity of carboxypeptidase can be illustrated as follows (*11*),

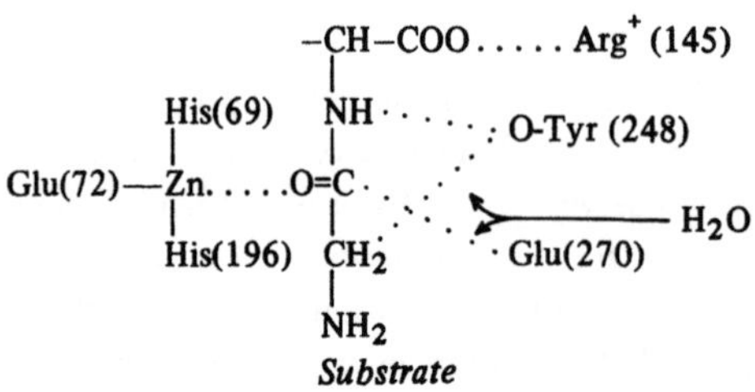

That is, two imidazoles (His) coordinate with Zn ion and control the electronic state of it. Though reaction of the enzyme to ester and peptide bonds are different, the role of Zn ion can be thought to be similar. The effective charge of Zn ion is reduced to +1 by Glu(72), and Zn ion coordinates with water when the substrate is absent. The position of the water is close to Glu(270), and they can interact with each other. The substrate expells the coordinated water giving the new interaction between Zn and O of carbonyl in the substrate. In this case, imidazole controls the coordinate activity of Zn ion, and then plays a cooperative role in the hydrolysis.

2.3. Hemochrome

The role of imidazole in hemochrome is more attractive. Structures of myoglobin (Mb), hemoglobin (Hb), cytochrome c(cytc) and cytocrome P 450 are illustrated in Fig. 1.

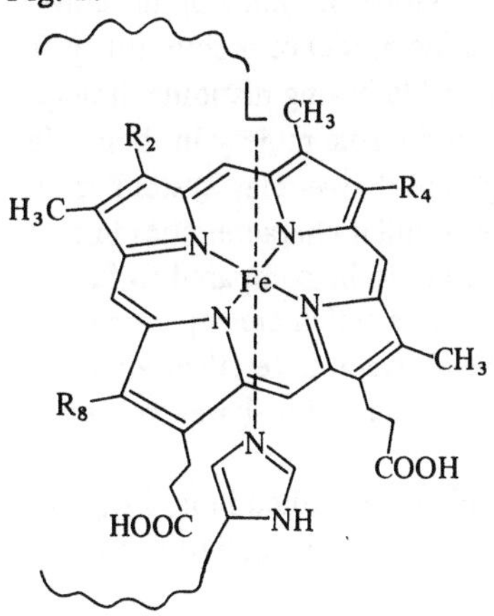

Hb, Mb

Peroxidase(horse radish) HRP
(R_2, R_4, $-CH=CH_2$: R_8, $-CH_3$)

cytc
(R_2, R_4, $-CH(CH_3)SCH_2CH(NH_2)COOH$; R_8, $-CH_3$)

cytc P450
(S^- as the legand in cytc)

Fig. 1. Structure of hemochrome

In the hemochrome imidazole moiety is responsible for the axial legand of the central metal ion. The function of the hemochrome is taken over by the metal, but a sufficient description has not yet been presented. The spin state of the metal ion provides important information about the activity of the hemochrome in which the axial legand affects the state by chelation. Table 2 shows the electronic structure of Hem iron.

Table 2. Electronic structure of hem iron

Oxidation state Spin state	Fe(II) $s = 2$ (high)	Fe(II) $s = 0$ (low)	Fe(III) $s = 5/2$ (high)	Fe(III) $s = 1/2$ (low)
Axial Legand (the sixth)	Mb^{II} Hb^{II} $P450^{II}$	$Mb^{II}(L)$ $Hb^{II}(L)$ $cytc^{II}$	$Mb^{III}(L)$ $Hb^{III}(L)$ HRP(L)	$Mb^{III}(L)$ $Hb^{III}(L)$ cytc(L) $P450^{III}(L)$
		L = O_2, CO, NO	L = H_2O, SCN^- OCN^-, F^- (OH^-)	L = CN^- (Im), (N_3^-) (Im derivative)

On the structure of HbO_2 two models are presented ($Fe(II)\text{-}O_2$, $Fe(III)\text{-}O_2^-$) (*12–14*). Iron in HbO_2 is in a low spin state. However, when imidazole coordinates as the sixth legand both high spin state and low spin state of Fe(III) are observed. These states relate to the interaction of oxygen with Hb, and analyses using NMR technique have recently been done (*15–17*).

3. The Cooperativity in the Catalysis by Nucleophile-Containing Polymers

α-Chymotrypsin is an enzyme having imidazole moiety of histidine at its active center and about this enzyme most information exists. Numerous nucleophile-containing polymers have been synthesized and studied as the model of α-chymotrypsin. As a model of α-chymotrypsin several kinds of oligopeptide having histidine moiety in their sequence (*18–20*), and vinyl polymers having imidazole moiety in their side chain (*21*) have been synthesized and studied. Though the charge-relay system cannot be observed in them, several cooperativities which are peculiar characteristics in the synthetic polymers are observed. These cooperativities are to be compared to functions of enzymes. Up to date we have few catalytic systems which are superior to the enzyme in a concrete form; on the other hand, the abstract materialization of enzymes with nucleophile-containing polymer has been done and it shares its own new field of chemical reaction.

In Table 3 vinyl polymers having nucleophile which were studied in hydrolysis of ester, are summarized focussing on the cooperativity in catalytic activity of the active center.

The catalysis is roughly divided into an adsorptive complex formation process and a reaction process. In kinetical treatment it is necessary to adopt an average profile, although the catalytic site in the nucleophile-containing polymer is multiform.

The kinetic scheme of the hydrolysis can be represented as follows:

$$\mathrm{Cat + S\text{-}S'} \underset{k_2}{\overset{k_1}{\rightleftharpoons}} \mathrm{Cat \cdot S\text{-}S'} \xrightarrow[\searrow \mathrm{S}]{k_a} \mathrm{Cat \cdot S'} \xrightarrow[\mathrm{H_2O}\nearrow]{k_d} \mathrm{Cat + S'}$$

(1) (2) (3)

$$1/k_{cat} = 1/k_a + 1/k_d, K = k_1/k_2$$

$$\nu = k_{cat} \cdot K \cdot [\mathrm{Cat}]\,[\mathrm{S\text{–}S'}]/[1 + K \cdot [\mathrm{S\text{-}S'}]]$$

where Cat is catalyst, S-S′ is substrate and ν is the rate of the hydrolysis. Step (1) is called Michaelis complex formation step, step (2) is the acylation step and step (3) is the deacylation step. Actually, the interaction between the catalyst and the substrate is not always the same as the Michaelis complex formation. Michaelis complex is an effective interaction between the catalyst and the substrate for the hydrolysis. The overall rate of the reaction depends on the magnitudes of K and k_{cat}. Accord-

Table 3. Catalytic activities of vinyl polymers containing nucleophile

---

Homopolymer

a Cooperativity between nucleophiles
IM(22, 23, 24, 25) BI (22, 23, 26)

b Less cooperativity between nucleophiles
NIM(22, 23, 27) 2 MIM(22) PY(41) TR(42) IR-51M(54, 55)

Copolymer

c Cooperativity between nucleophile and binding moiety
IM-ph(28) IM-la(29) IM-ac(30, 31, 32)
API-mac(33 API-ac(34, 35) API-aba-pr(33) API-mc(33)
2 MIM-2nmim(34, 35)

d Effective binding for the catalysis
IM-al(28)
2MIM-pr(36) 2MIM-am(36) 2MIM-ac(34, 35)
ABI-pr(37) ABI-am(37) ABI-ap-am(37)
API-pr(34, 35) API-am(37) API-ap-am(37) API-anp-am(33) API-aha-am(33)
PY-apy(21)
PH-3mim(43)

e Non-effective binding for the catalysis
IM-pr(27) IM-sc(31)
NIM-pr(27)

f Molecular relay system
PHA-IM-am(38) MHA-2MIM-am(39) PHA-2MIM-am(40)

Nucleophile (catalyst moiety)*

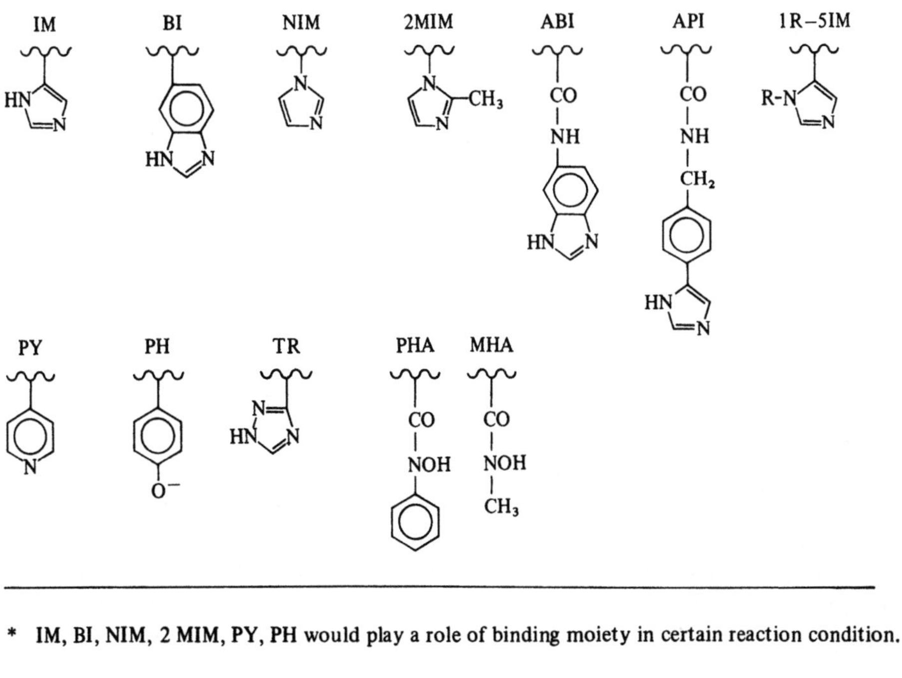

---

* IM, BI, NIM, 2 MIM, PY, PH would play a role of binding moiety in certain reaction condition.

Table 3. (continued)

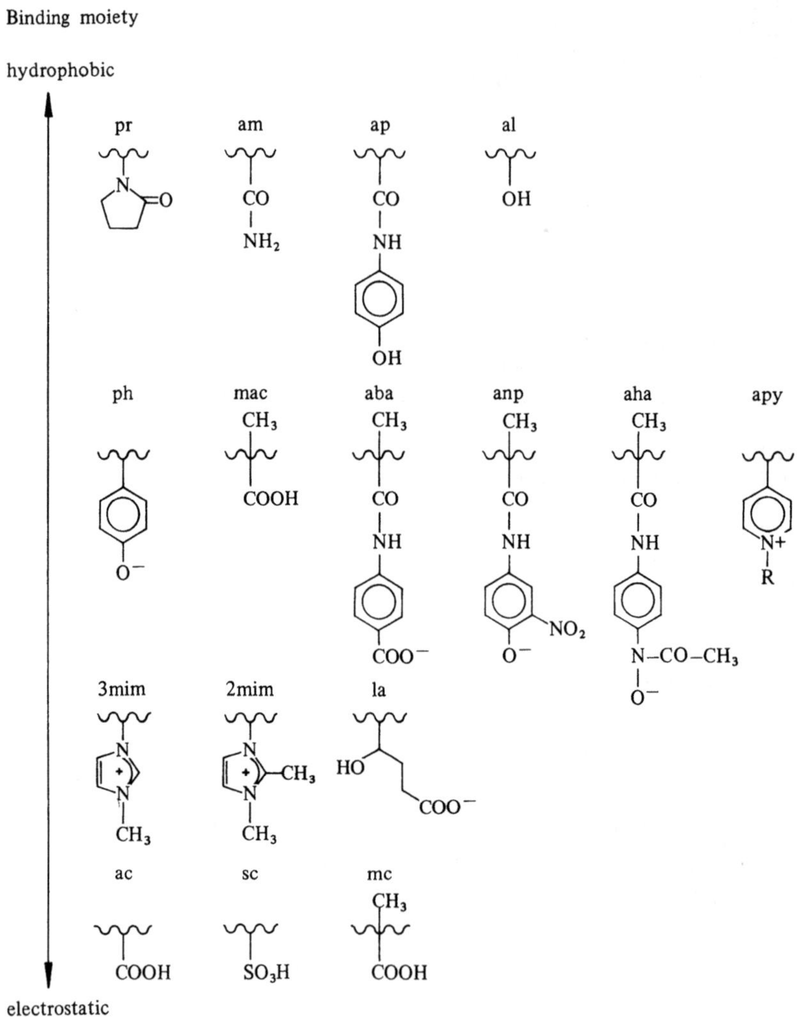

ingly, to discuss the catalytic activity, a kinetical analysis which contributes to the overall rate is necessary.

Two expressions for the pre-equilibrium step in the reaction as in the above equation can be adopted; one is the Michaelis-Menten equation (*44*) and the other is the Langmuir isotherm (*45*). Philosophically, the former is treated for a continuous reaction and the latter is done for a reaction whose steps can be analyzed independently. For the reaction system in which quantities of the adsorptive site and the catalytic site vary, the kinetical treatment by the latter is convenient. In this article, both treatments are adopted. Generally, desorption is faster than the acylation ($k_2 \gg k_a$), so that the following equation can be expressed as

$$K_m \cong K^{-1}$$

where K_m is Michaelis constant and K is equilibrium constant of the effective adsorption (the effective binding) of the substrate on the catalyst.

In the case of less contribution of the pre-equilibrium step to the reaction ($K \cong 0$), the rate can be expressed as

$$\begin{aligned} v &= k_a \cdot k_d \cdot [\text{Cat}]\,[\text{S–S}']/[k_a[\text{S-S}'] + k_d] \\ &= k_{cat} \cdot [\text{Cat}]\,[\text{S-S}'] \end{aligned}$$

3.1. Binding

The effective binding of the substrate for the catalysis plays an important role in giving a high catalytic activity as was described above. The binding of the substrate to the nucleophile-containing polymer is not as specific as compared with that of the enzyme. However, some nucleophile-containing polymers have binding ability as high as α-chymotrypsin in magnitude and thermodynamical free energy change, when a coulombic interaction force is taking place in the binding.

The binding forces are as follows: coulombic interaction, hydrophobic interaction, hydrogen bond formation and charge transfer interaction. Each force is given in the correlation of the polymer and the substrate.

The coulombic interaction is shown in the equation of

$$E(r) = \sum_i \sum_j (q_i \cdot q_j / D \cdot r_i \cdot r_j)$$

where q_i and q_j are charges, r_i and r_j are distances between the charge points, and D is the dielectric constant of the medium, respectively. The dielectric constant affects the force substantially and it is usually very large in the polymer domain.

The hydrophobic interaction is estimated indirectly with changes in the water structure and in the free energy of the system. For the polymeric system the interaction force is given in the order of several kcal.

The hydrogen bond and the charge transfer interaction create various forces according to the structures of the bindings.

In any case, the most significant requirement for the binding for the catalysis is that the binding takes place in a favorable position for the reaction. The magnitude of the interaction and that of the effective binding do not always coincide.

The Binding by Coulombic Interaction

The coulombic force is a long-range one. The force is very effective for the binding.

Studies of hydrolyses using the coulombic interaction between polymeric catalyst and charged substrate have been carried out by Letsinger (*41*) and Overberger (*23*) in previous years. The rate of the hydrolysis of the substrate of an opposite charge to that of the charged groups on the polymer, such systems as poly(4-vinylpyridine), PVP (I), – NABS(1), – NAS(2), – NABA(3) and poly(4(5)-vinylimidazole), PVI (II), – NABA(3), reaches the optimum in the reaction condition of that 20–30% of the nucleophile moiety is protonated in certain pH of the reaction solution. These results were considered as showing that the coulombic interaction between

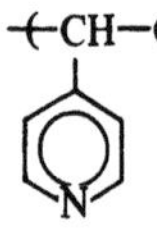
(I)

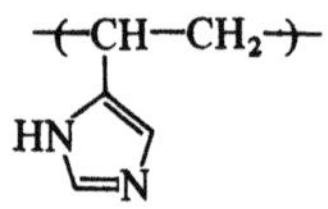
(II)

NABS

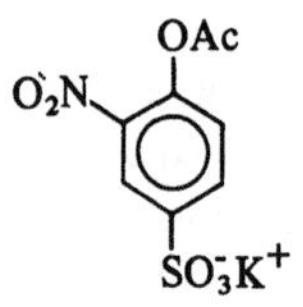
(1)

NAS

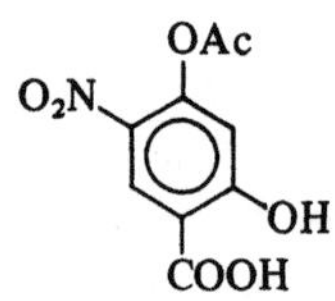
(2)

NABA

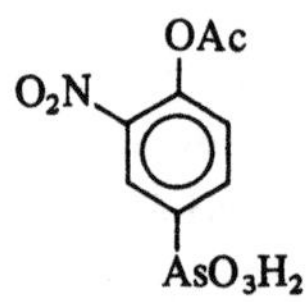
(3)

the catalyst and the substrate increases the rate of the hydrolysis. In the case of the hydrolysis of the anionic polymer substrate (4), by partially protonated poly(N-vinylimidazole) (III), the rate is higher than that by N-methylimidazole, monomeric analogue of III, and shows a Michaelis-Menten profile (*46, 47*). The Michaelis constant, K_m, is estimated to be 2.4×10^{-5}M ($K \cong 4.2 \times 10^4 M^{-1}$).

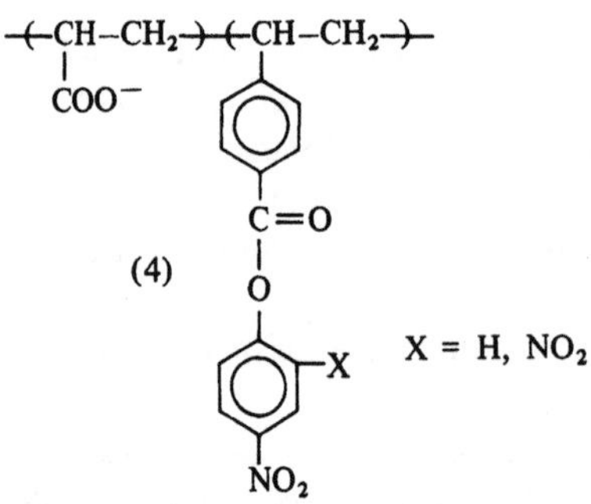
(4)

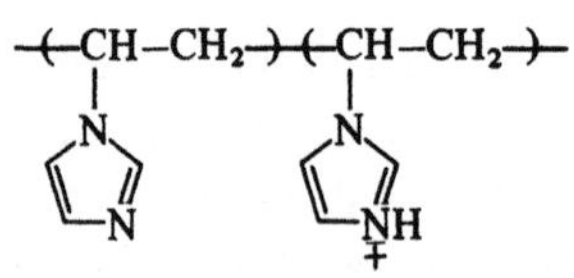
(III)

PNPA

(5)

ANTI

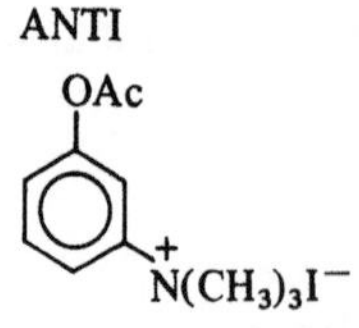
(6)

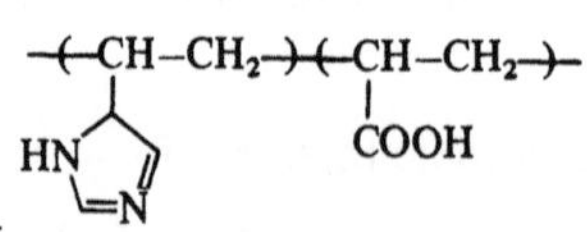
(IV)

The fact that poly(acrylic acid) inhibits the reaction and that the neutral substrate PNPA(5), is not bound to the polymer, show the significance of the coulombic binding in the reaction.

Poly(4(5)-vinylimidazole-co-acrylic acid), Poly(IM-ac) (IV), which contains anionic binding sites also shows the significance of the binding. The overall hydrolysis rate increases in the order of the positively charged ANTI(6), the neutral PNPA(5), and the negatively charged NABA(3), adopted as the substrates. The hydrolyses of ANTI(3) by monomeric analogues such as imidazole and imidazolylpropionic acid are faster than the hydrolysis by Poly(IM-ac) (IV) (*30*). On the other hand, the catalytic activity of the imidazole-sulfonic acid copolymer is very low. This is explained by assuming inertness of the pendent neutral imidazole moiety by the destruction of a zwitterion and repelling hydroxide ion which is considered as the hydrolytic reactant (*31*). In this case, an unfit binding would be considered.

The significance of the effective binding is proved in the same catalytic system (*32*). The value of the effective binding for the hydrolysis, the reciprocal of Michaelis constant, is not proportional to the amount of carboxylic acid moiety in Poly(IM-ac), and gives the optimum at about 30 mol% of IM content in the polymer. The most preferable carboxylic acid (ac) content in the sequence of the polymer for the hydrolysis reaction is about 70 mol%. (Fig. 2). This shows that the effective binding

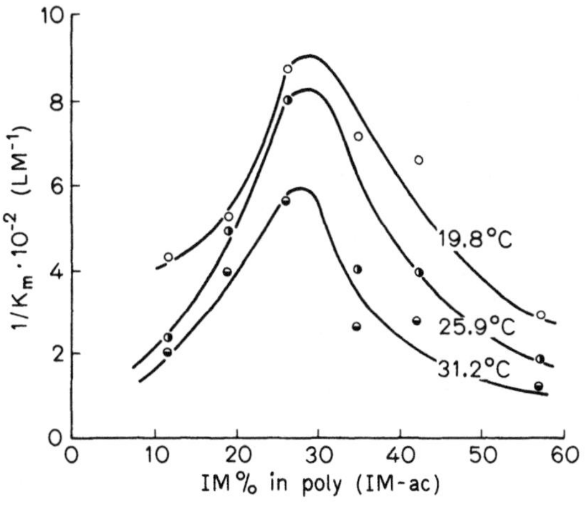

Fig. 2. Plots of reciprocal Michaelis Menten constant against copolymer composition in poly (IM-ac) (pH = 9.0)

for the hydrolysis is affected by carboxylic acid moiety adjacent to the IM in the polymer. The effective binding decreases with temperature, but the optimum point is not shifted. At the optimum point drastic changes are observed in ΔH and ΔS in the binding where a compensative relationship between them is maintained.

In oxidation reactions of ascorbic acid, homogentisic acid and hydroquinone by poly(l-histidine) – Cu(II) complex, the reaction profile shows a Michaelis-Menten type curve in the reaction condition of pH ranging from 5–6. These oxidation reaction rates are higher than the oxidation rates by the catalyst without poly(l-histidine). On the other hand, the rate of oxidation of a positively charged substrate, phenylenediamine, by the complex, is lower than that by the catalyst without poly(l-histidine). In these cases the interaction between the positively charged imidazole and the substrates is also indicated (*48*).

The Binding by Hydrophobic Interaction

The hydrophobic interaction is a relatively short range interaction, and it is not so distinct compared with the coulombic interaction. A favorable binding by the hydrophobic interaction also contributes to increase the overall rate of the catalysis. Many polymers which contain hydrophobic groups as the binding moiety have been synthesized and studied.

The hydrolysis of NABA(3) by poly(2-methyl-4(5)-vinylimidazole-co-N-vinylpyrrolidone), poly(2MIM-pr), gives a Michaelis-Menten type kinetics (*36*). On the other hand, the hydrolysis of NABA by poly(2-methyl-4(5)-vinylimidazole-co-acrylamide), poly(2MIM-am), does not show a Michaelis-Menten kinetics. The difference between them is due to the binding of the substrate and the polymer. Results are shown in Table 4. The binding is thought to be a hydrophobic one, because pyrrolidone has the nature of the hydrophobic interaction (*50*) and acrylamide moiety has that of a less hydrophobic one. Additions of benzyl alcohol and dioxane to the reaction system cause a competitive inhibition in the hydrolysis. This also implies that the binding is a hydrophobic one (*49*).

Table 4. Hydrolyses of NABA by poly(2MIM-pr) and poly(2MIM-am)

Polymer	2MIM content (mol %)	K_m (M)	k_{cat} (min^{-1})
Poly(2MIM-pr)	3.5	$8.7x10^{-3}$	$3.6x10^{-2}$
	7.5	$8.8x10^{-3}$	$4.2x10^{-2}$
	15.7	$10.9x10^{-3}$	$3.6x10^{-2}$
Poly(2MIM-am)	2.8	$66.0x10^{-3}$	$11\ x10^{-2}$

pH = 8.0; 30 °C; μ = 1.0 (KCl).

Existence of pyrrolidone moiety in the polymer does not always assist in the favorable hydrophobic binding. Poly(N-vinylimidazole-co-N-vinylpyrrolidone), poly(NIM-pr), gives the second order kinetics in the hydrolysis of PNPA(5).

Both benzimidazole-containing polymer and phenylimidazole-containing polymer show the hydrophobic binding in the catalyses, obviously (*35, 37*). The polymers are poly(ABI-pr), poly-(ABI-am), poly(ABI-am-ap), poly(API-pr), poly(API-am) and poly(API-am-ap). Imidazole moieties in these polymers have hydrophobic binding properties themselves. Profiles of hydrolyses of p-acetoxybenzoic acid, ABA(7), show the Michaelis-Menten kinetics. Catalytic activities of those polymers are more than 30 times that of low molecular weight analogues. Several sets of data are tabulated in Table 5.

ABA

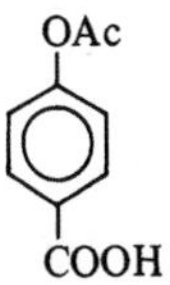

(7)

Table 5. Hydrolyses of ABA by several polymers containing hydrophobic binding moiety

Polymer	Cont. of IM moiet. (mol%)	K_m (M)	k_{cat} (min^{-1})	n' [a]
Poly(ABI-pr)	0.91	2.3×10^{-2}	9.2×10^{-3}	0.98
	1.22	2.4×10^{-2}	9.2×10^{-3}	1.03
	2.90	2.4×10^{-2}	9.7×10^{-3}	1.01
Poly(ABI-am)	1.29	5.3×10^{-2}	1.9×10^{-2}	1.01
	7.83	7.7×10^{-2}	1.8×10^{-2}	1.56
Poly(ABI-am-ap)	3.2 (ap = 3.4)	3.1×10^{-2}	5×10^{-3}	1.06
	3.5 (ap = 15)	4.0×10^{-2}	5.3×10^{-3}	1.10
Poly(API-pr)	1.9	1.9×10^{-2}	1.3×10^{-2}	0.98
Poly(API-am)	1.68	8.3×10^{-2}	1.9×10^{-2}	1.09
	8.91	2.9×10^{-2}	2.9×10^{-2}	1.24
Poly(API-am-ap)	1.7 (ap = 5.0)	1.2×10^{-2}	1.7×10^{-2}	1.14

pH = 8.0, 30 °C, μ = 1.0 (KCl).
[a] $pK_a = pH + n'\log[(1-\alpha)/\alpha]$
which relates to electrostatic interaction

In poly(ABI-pr) no effect of pyrrolidone moiety on the binding and the catalysis is observed. On the contrary, acrylamide moiety affects the binding in poly(ABI-am). This is explained as a considerable interaction of benzimidazoles themselves taking place in poly(ABI-am). In addition, that the rate constant of the catalysis, k_{cat}, of poly(ABI-am) is higher than that of poly(ABI-pr) is explained by a dynamical binding which covers the lower binding expressed in K_m. The addition of phenol moiety in the poly(ABI-am) sequence, which is represented as poly(ABI-am-ap), increases the binding but decreases the catalysis. Thus the phenol moiety would contribute to an increase in the hydrophobicity of the catalytic center. In the case of the phenyl-imidazole-containing polymers similar arguments are given. On these discussions the schematic representation as shown in Fig. 3 is given (*37*).

Table 6 shows thermodynamic parameters of the binding processes. In spite of the polymeric catalyst and the low molecular weight analogues the similar correlationship between ΔH and ΔS_u is given, where unitary entropy changes, ΔS_u, as

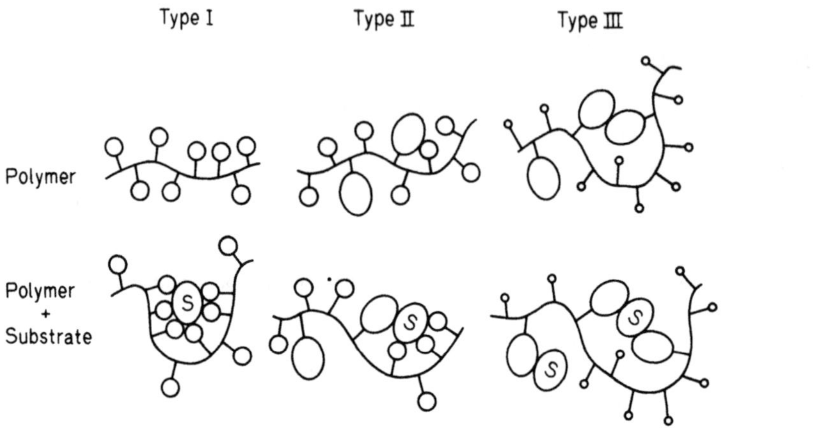

Fig. 3. Schematic representation of the mode of intramolecular aggregation and substrate binding
Type I: poly(2MIM-pr)
Type II: poly(ABI-pr), poly(API-pr)
Type III: poly(ABI-am), poly(ABI-am-ap), poly(API-am) poly(API-am-ap)
From J. Amer. Chem. Soc., *93*, 4252 (1971)

Table 6. Thermodynamic parameters for the binding of ABA

Polymer	ΔH (kcal · M^{-1})	ΔS (cal · M^{-1} · T^{-1})	ΔS_u (cal · M^{-1} · T^{-1})	ΔG_u(303 °K) (kcal · M^{-1})
Poly(ABI-py) (Im = 1.94%)	1.20	11.8	19.8	–4.80
Poly(ABI-am) (Im = 8.91%)	5.0	24.9	32.9	–4.57
Low mol.analog.				
(I)	3.74	20.5	28.5	–4.90
(II)	3.44	19.9	27.9	–5.01

(I)

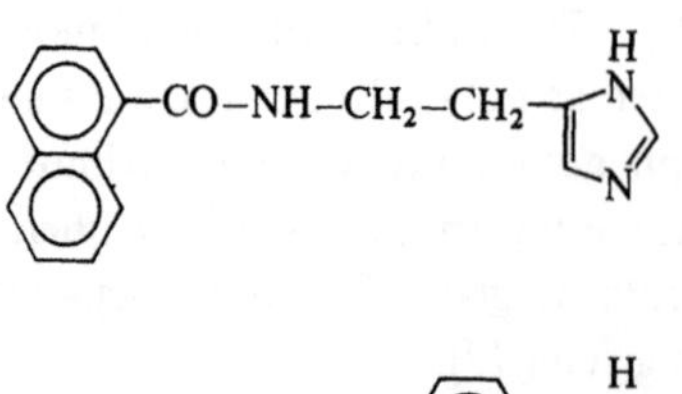

(II)

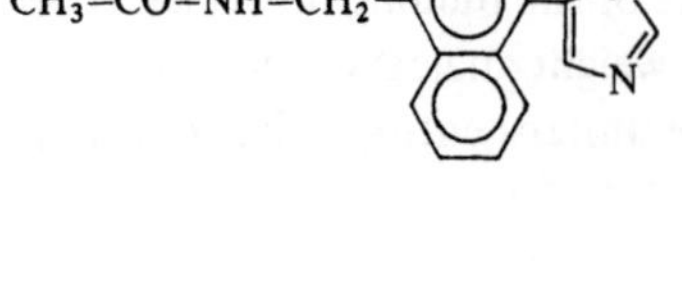

estimated by the equation of $\Delta G_u = \Delta G - 7.98\ T$ (*52*). From these data, the iso-equilibrium temperature is determined to be 320 ± 20 °K. This temperature is similar to that obtained in the hydrophobic binding of organic molecules with poly(N-vinyl-pyrrolidone) (*50*). Accordingly, this temperature is considered to indicate the melting point of an iceberg, and the dehydration of imidazole moiety by the melting of the iceberg and the hydrophobic binding of the substrate is suggested (*35*).

The increase of hydrophobicity in the substrate shows the increase in the binding. The hydrolysis of 3-nitro-4-acyloxy benzoic acid (8) by poly(4(5)-vinyl-imidazole) (II) is much influenced by the acyl chain length in the substrate (*24, 25, 56, 72*). The increase of the acyl chain length makes the binding increase and the deacylation decrease. In the hydrolysis of the substrate having the longer acyl chain, a micelle effect should be considered. The hydrolysis is much influenced by the micelle formation.

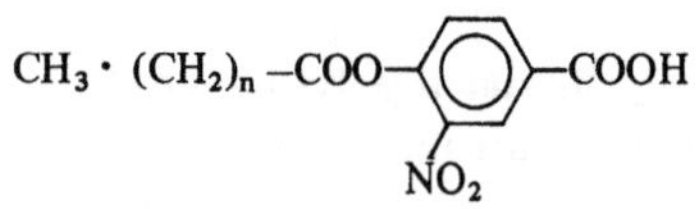

(8)

Both poly(l-butyl-5-vinylimidazole), poly(1B-5IM), and poly(1-methyl-5-vinyl-imidazole), poly(1M-5IM), have no quaternizable nitrogen, so that the coulombic interaction with the substrate is not necessarily to be considered. Using these polymers, hydrolyses of several 3-nitro-4-acyloxy benzoic acid having various acyl chain length (8) (n = 0, 5, 10, 16) have been made (*55*). The apparent rates of hydrolyses are tabulated in Table 7.

Table 7. Hydrolysis rates of 3-nitro-4-acyloxy benzoic acid (8) by poly(1B-5IM) and poly(1M-5IM)

Catalyst	S2	S7	S12	S18
1,5-Dimethylimidazole	8.93	4.12	3.64	0.957
Poly(lM-5IM)	12.8	7.88	11.15	85.0
Poly(lB-5IM)	7.78	0.78	64.10	77.2
				($\times 10^3 min^{-1}$)

27.6% ethanol-3.3%acetonitrile-69.1% water; pH = 6.85; μ = 0.02; 26 °C; [Cat] = 5×10^{-4}M; [Sn] = 5×10^{-5}M.

In general the hydrolysis rate increases with the acyl chain length in the substrate, Sn (8), and polymeric catalysts have high activities compared with the low molecular weight analogue. However, there are some exceptions. The longer acyl chain in the substrate does not always lead the higher hydrolysis rate. The hydrolysis rate of S 12 by poly(lm-5IM) is lower than that by poly(lB-5IM), conversely, the rate of S 18 by poly(lm-5IM) is higher than that by poly(iB-5IM). This is explained by a steric hindrance with the bulky butyl group in the polymer. Still, some problems exist in these catalyses. The hydrolysis of S 18 by poly(lM-5IM) shows a Michaelis-Menten kinetics, besides, that by poly(lB-5IM) shows the second order kinetics contrary to expectation of the larger hydrophobic interaction between the polymer and the substrate. This also implies the significance of the favorable binding.

The effect of the hydrophobicity of the substrate can be seen elsewhere. Partially benzylated polyethyleneimine binds p-nitrophenyl ester having long chain alkyl groups in the hydrolysis (*53*). The binding is considered to be hydrophobic interaction. The concept of the benzylated N atom in the polymer acting as a general base catalyst is supported by studies of isotope effect and spectroscopy on this catalytic system. In addition, the hydrolyses of esters by polymeric sulfonic acid (*58–60*) and by long chain alkylated polyethyleneimine (*61, 62*) are shown as examples. Also, hydrolyses of the substrate (8) by several kinds of imidazole-containing polymers have been carried out and the contribution of the hydrophobic binding to the catalysis is indicated (*72*).

The hydrolyses of the neutral substrates by partially alkylated poly(4-vinylpyridine) (I) were studied (*63, 64*). The rate enhancement is attributed to the hydrophobic interaction between the substrate and the polymer. In this hydrolysis, the rate of the substrate decomposition and the rate of the catalysis hold almost constant. It was explained that the deacylation step in the hydrolysis is affected by the bulkiness of the alkyl group in the polymer in which the alkyl group interferes an attack of water molecule on the acylated intermediate.

Several histidine-containing cyclic dipeptides catalyse phenolic acetates (*65*). Cyclo(Gly-L-His), cyclo(L-Leu-L-His) and cyclo(D-Leu-L-His) have almost the same pK value of their imidazole moieties as that of imidazole. Nevertheless, catalytic activities of the former two cyclic dipeptides are less than that of imidazole and the latter has an activity of 20-times that of imidazole. This was also explained by showing that there are considerable steric hindrances influencing the catalysis and that the latter has a remarkably effective hydrophobic interaction which exceeds the hindrance.

The Binding by Coulombic and Hydrophobic Interaction

Many polymer catalysts whose binding sites have both coulombic and hydrophobic interactive property have been synthesized. It cannot be stated conclusively which property is dominant.

In the comparison of hydrolyses of positively charged substrates such as ANTI(6) and NNTI(9) with negatively charged polymer catalysts such as poly(API-ac) and poly(API-mc) a large binding constant is given in the hydrolysis of the NNTI (*33*). The binding of NNTI to poly(API-ac) is larger than that to poly(API-mc). That poly(API-mc) has a lower binding property with NNTI was also explained by the polymer conformation as is stated above in Fig. 3 (*33*).

In the case of the hydrolysis of the negatively charged substrate, NABA(3), by partially N-methylated poly(N-vinyl-2-methylimidazole) (V) the decrease of the degree of the N-methylation increases both the binding constant, K, and the rate

NNTI

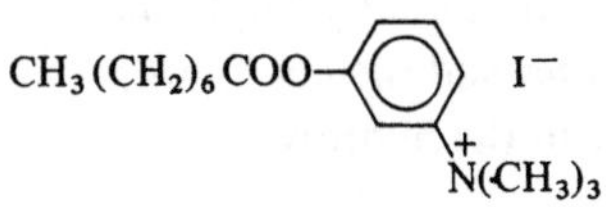

(9)

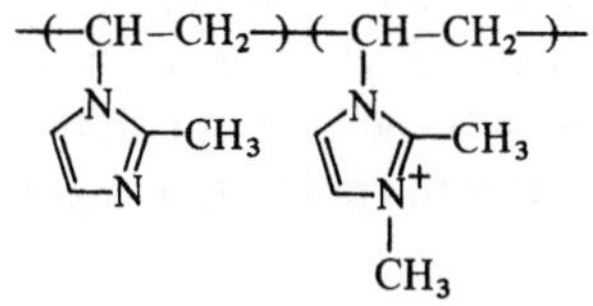

(V)

constant of the catalysis, k_{cat} (*66*). This suggests that both the hydrophobic and the coulombic interactions take place in the hydrolysis. When acrylamide or N-vinyl-pyrrolidone is introduced as the comonomer into the polymer the binding constant and the catalytic rate constant vary reversely. This result leads to the conclusion that the hydrophobic property takes place in preference to the coulombic property in the binding in this case.

A combined catalyst such as imidazole-containing polymer and imidazole acylated by a long chain acyl group gives a similar result in the hydrolysis of 3-nitro-4-acyloxy benzoic acid (8) (*72*).

As the hydrophobic binding is influenced by the reaction solution the hydrolysis has been carried out in several reaction solutions. The increase of the ethanol content in its aqueous solution decreases the magnitude of the hydrophobic interaction. Table 8 shows the result of the hydrolysis of the negatively charged substrate, NDBA [n = 10 in (8)], by poly(4(5)-vinylimidazole) (II) in an aqueous ethanol solution (*24*).

Table 8. The hydrolysis of NDBA by poly(4(5)-vinylimidazole) in ethanol aqueous solution

Ethanol cont. (%)	pH	K (M^{-1})	k_{cat} (min^{-1})
20	7.90	$2.6x10^4$	11.4
30	7.90	$3.2x10^3$	20.7
43.7	7.99	$2.1x10^3$	0.032

μ = 0.02; 26 °C.

The binding constant, K, decreases with ethanol content in the reaction solution and the rate constant of the catalysis gives the optimum at about 30% of ethanol. The binding constant also gives the optimum at a temperature of 25 °C. These results are interesting in connection with the structure of water which gives the optimum at the condition of 27% of ethanol content and 25 °C (*24, 25*).

Contributions of both the hydrophobic and the coulombic interaction are indicated in the hydrolyses of 3-nitroacyloxy benzoic acid (8) by poly(5(6)-vinyl-benzimidazole) (*67*) and of 3-acyloxy-N-trimethylanilinium iodide (acyloxy group in lieu of acetoxy group in ANTI (6)) by poly(5(6)-vinylbenzimidazole-co-acrylic acid), poly(BI-ac) (*68*).

Such a contribution is also observed explicitly in the hydrolyses of esters by polycations containing various alkyl groups (*69*). The fact that the hydrophobic binding of the substrate with the alkyl group in the polymer and the coulombic interaction between OH^- and the polycation takes place in the hydrolysis was concluded.

The Binding by Hydrogen bond Formation

N-Carboxyanhydride of α-amino acid is polymerized by the nucleophile-containing polymers such as poly(N-vinyl-2-ethylimidazole), poly(N-vinyl-2-ethylimidazole-co-acrylamide) and poly(N-vinyl-2-ethylimidazole-co-N-vinylpyrrolidone) (*102, 103*). The rate of polymerization increases with the order of pyrrolidone, amide and imidazole moieties. The order corresponds to the magnitude of the bond formation which is observed in IR measurement. The rate also increases with the content of pyrrolidone moiety. The rate in N,N'-dimethylformamide solution which is unfavorable to the hydrogen bond formation is lower than that in acetophenone. These results show that the hydrogen bond formation is notable as the binding.

The Binding by Charge Transfer Complex Formation

The binding by the charge transfer complex formation is observed in polymerizations of p-nitrophenylphenylalanine NCA and some p-nitrophenyl derivatives of α-amino acid NCA by the nucleophile-containing polymer catalysts (*103*). Results are shown in Table 9.

Table 9. Rates of polymerization of phenylalanine NCA and its derivatives by some imidazole-containing polymers

Polymer	Polymerization rate of NCA (min^{-1})		
	DL-β-Phe	p-Nitro-DL-β-Phe	o, p-Dinitro-DL-β-Phe
Poly(N-vinyl-2-ethylimidazole)	$2.7 x 10^{-4}$	$1.1 x 10^{-3}$	$2.1 x 10^{-3}$
Poly(N-vinyl-2-ethylimidazole-co-N,N'-diethylacrylamide) (34:66)	$3.0 x 10^{-3}$	$1.2 x 10^{-2}$	–
Poly(N-vinyl-2-ethylimidazole-co-N-vinylpyrrolidone) (30:70)	$4.8 x 10^{-3}$	$4.8 x 10^{-2}$	$7.0 x 10^{-2}$

Solvent, acetophenone; 35 °C

The effect of the hydrogen bond formation is much greater than that of the charge transfer complex formation in this polymerization reaction.

3.2. Activity of the Catalytic Site

Many cooperative interactions can be found between pendant functional groups in the nucleophile-containing polymers. Features of the cooperations are full of variety. The cooperative interaction is shown in poly(4(5)-vinylimidazole), by Overberger (*70*), in the earlier studies.

Cooperative Interaction Between Imidazole Moieties

Imidazole provides three chemical species according to pH of solution. That is to say, the cationic (IM^+), the neutral (IM), and the anionic (IM^-) imidazole.

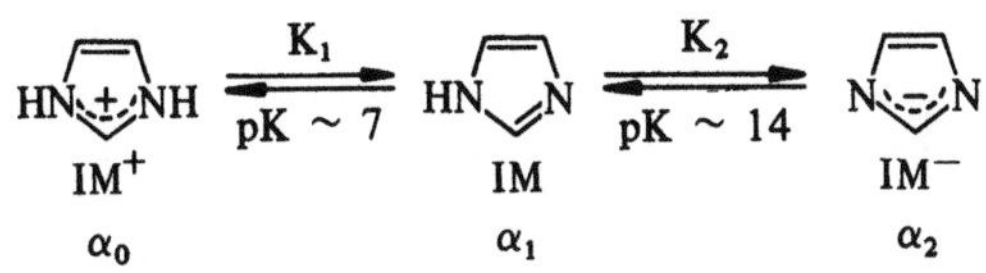

Accordingly, several kinds of combinations such as IM^+-IM^+, IM^+-IM, IM-IM, IM-IM^-, IM^--IM^-, can be thought to be in poly(4(5)-vinylimidazole). The cooperation of imidazole in the catalysis can be seen in the hydrolysis of sym-dichloroacetone by imidazole in 95% dioxane solution (*71*). The rate of the hydrolysis is represented as

$$v = k_0 + k_1[\mathrm{IM}] + k_2[\mathrm{IM}]^2$$

The third term shows the cooperation of imidazole moieties. In the hydrolysis by tertiary amine no dependence of the square of the amine concentration is shown as the third term in the above equation. An electron transfer as in the following scheme is considered,

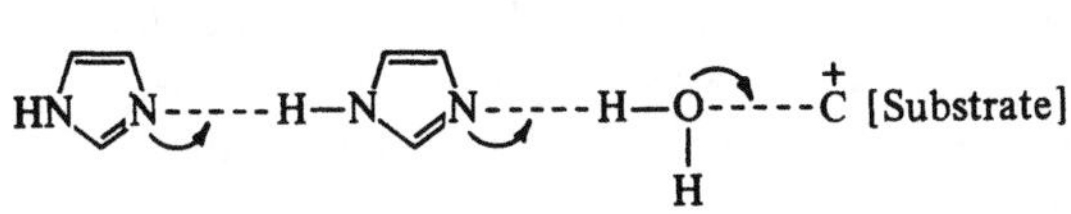

That is, the activated imidazole by the imidazole on the left hand side activates the water as a general base. This is considered to be a stabilization of the imidazolium ion.

In poly(4(5)-vinylimidazole) notable phenomena related to the cooperation can be seen.

The degree of polymerization of poly(4(5)-vinylimidazole) greatly affects the catalysis rate, k_{cat}, and the fraction of the neutral imidazole moiety in the polymer (*73, 90*). In hydrolyses of the negatively charged substrate, NABS (1), and the neutral substrate, PNPA(5), the k_{cat} increases with the degree of polymerization. The increase in the degree of polymerization causes the higher possibility of the cooperation of imidazole moieties in the polymer.

IM⁺-IM cooperation. The evidence of the cooperation between IM^+ and IM lower pH range is less than that of imidazole. This means that there is no considerable cooperation between IM^+ and IM^+.

IM^+-IM cooperation. The evidence of the cooperation between IM^+ and IM moieties is not given. The role of IM^+ is considered to be the binding site and gives a positive electric field to the catalytic site.

IM-IM cooperation. This cooperation can be found in the hydrolyses of the neutral substrate, PNPA(5), by poly(4(5)-vinylimidazole) and of the negatively charged substrate, NABS(1), by poly(5(6)-vinylbenzimidazole). These hydrolyses are achieved in an intermediate pH range (*22, 26*). Figure 4 shows the hydrolyses

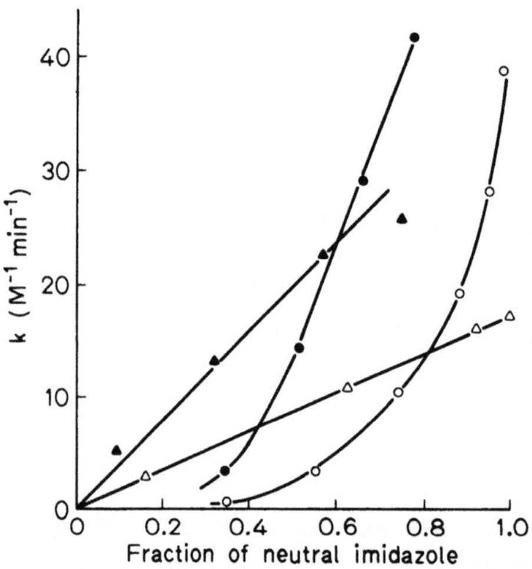

Fig. 4. Rates of hydrolysis of PNPA by poly(4(5)-vinylimidazole) and imidazole
10% methanol-water system
● poly(4(5)-vinylimidazole)
▲ imidazole
28.5% ethanol-water-system
○ poly(4(5)-vinylimidazole)
△ imidazole

rates of PNPA by poly (4(5)-vinylimidiazole) in 10% methanol-water and 28.5% ethanol-water systems. Both rates increase exponentially with the fraction of the neutral imidazole moiety. On the other hand, the rates of hydrolyses by imidazole increase linearly with the neutral imidazole fraction. The enhancement of the hydrolysis rate in a 10% methanol-water system is larger than that in a 28.5% ethanol-water system. This fact indicates that the hydrophobic binding may take place largely in a solvent of higher polarity. As a result, the exponential increase in the hydrolysis rate by the polymer correlates to the cooperation of imidazole moieties in the polymer. The following two schematic mechanisms (*22*) are considered.

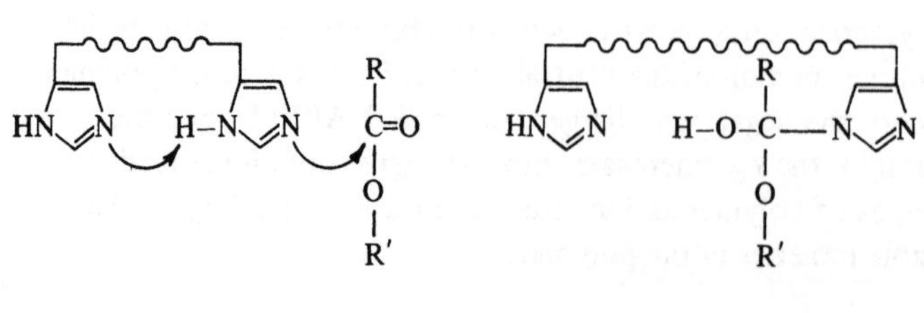

Poly(1-methyl-5-vinylimidazole), poly(1M-5IM), does not show such cooperation (*54*). The hydrolysis rate of PNPA(5) by poly(1M–5IM) is about a fifth of that by 1,5-dimethylimidazole.

For the cooperation a consideration of the contribution of tautomeric imidazole for the catalysis was also presented (*73*).

The reduced activity of poly(4(5)-vinylimidazole) in comparison to imidazole in the hydrolyses at low α_1 values, as is shown in Fig. 4, is considered (*70*): A larger fraction of the protonated imidazole moieties excludes the substrate or a more extended conformation of the polymer which is caused by an increase of the protonated imidazole moieties and renders difficult the IM-IM cooperation.

Table 10 shows the differences of thermodynamic parameters in the hydrolyses of PNPA by poly(4(5)-vinylimidazole) and poly(5(6)-vinylbenzimidazole) in comparison with their low molecular weight analogues, in the range of pH 8.0–8.4. This implies that the entropic factor is considerable.

Table 10. Differences of thermodynamic parameters in the hydrolyses of PNPA by poly(4(5)-vinylimidazole) and poly(5(6)-vinylbenzimidazole) and their low molecular weight analogues

Catalyst Polymer	Low mol. analog.	$\Delta(\Delta H^{\ddagger})$ (kcal·M^{-1})	$\Delta(\Delta S^{\ddagger})$ (cal·M^{-1} · T^{-1})
Poly(IM)	Imidazole	−3.3	−10
Poly(BI)	Benzimidazole	−2.8	− 6

pH = 8.0 ~ 8.4.

IM-IM⁻ cooperation. It is difficult to establish the IM-IM⁻ cooperation in poly(4(5)-vinylimidazole), since pK_2 value is more than 14.5 (*74*). The IM-IM⁻ cooperation can be observed in hydrolyses of PNPA(5) and 4-acetoxy-3-nitrobenzoic acid [n = 0 in (8)] by poly(5(6)-vinylbenzimidazole), of which pK_2 value is estimated to be 12.7 (*26*). The hydrolyses rates increase sharply as the pH approaches the pK_2 value of the polymer. In contrast, such a drastic enhancement in the rate is not observed in the benzimidazole-catalysed hydrolysis, nevertheless, pK_2 value of the benzimidazole being 12.7. N-Alkylated polymers such as poly(N-vinylimidazole), poly(NIM), and poly(2-methyl-N-vinylimidazole), poly(2MIM), which have no anionic form in imidazole moieties, show the hydrolytic activities slightly less than that of poly(4(5)-vinylimidazole). In cases of low molecular weight analogues, such as N-methylimidazole and imidazole, the hydrolytic activity of the former is about 75% of the activity of imidazole (*75, 76*). These phenomena appear to support the IM-IM⁻ cooperation in poly(4(5)-vinylimidazole) indirectly. In the hydrolyses of NABS(1) and NABA(3) by poly(5(6)-vinylbenzimidazole), the activity enhancement is about 50-fold compared with the activity of benzimidazole, in spite of these anionic substrates being used (*23, 26*).

IM⁻-IM⁻ cooperation. Concrete evidence of the IM⁻-IM⁻ cooperation is not obtained.

IM-OH cooperation. It is interesting that the appearance of a cooperation between imidazole and hydroxyl moieties in the synthetic polymer is found, as is observed in α-chymotrypsin-catalyzed hydrolysis. The cooperation is ascertained in the hydrolyses catalyzed by copolymers of 4(5)-vinylimidazole with vinyl alcohol, poly(IM-al) (VI), and with p-vinylphenol, poly(IM-ph) (VII) (*77, 78*).

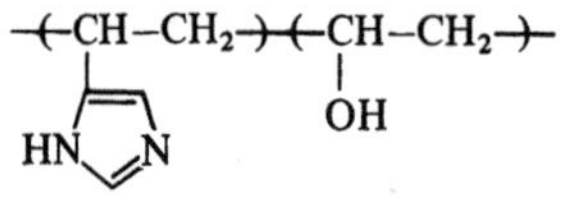

(VI)

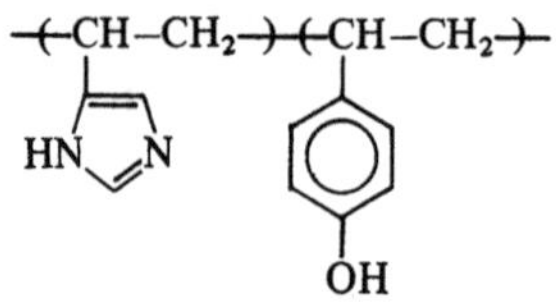

(VII)

Both polymers enhance the hydrolysis of PNPA(5) in a higher pH range. In them poly(IM-ph) enhances the hydrolysis remarkably. Meanwhile, poly(vinyl alcohol), poly(p-vinylphenol) and copolymer of 4(5)-vinylimidazole and p-methoxystyrene have little effect on the hydrolyses. These results indicate the cooperation of $IM\text{-}O^-$, especially of imidazole – phenolate anion. The molecularity of the cooperation is not determined but it is suggested that the molecularity of the cooperation in poly(IM-al) is not so simple. In the case of the hydrolysis of the negatively charged substrate, NABS(1), by poly(IM-ph), the hydrolysis rate in the higher pH range ($\alpha_1 > \sim 0.8$) increases remarkably, compared with that by poly(4(5)-vinylimidazole) (*77*). This also shows the significance of the imidazole-phenolate anion cooperation as well as the significance of the contribution of the anionic substrate (a concerted attack from both sides) in the catalysis.

IM-COOH cooperation. Copolymers of 4(5)-vinylimidazole and acrylic acid, poly(IM-ac), are considered as the His-Asp cooperation model in α-chymotrypsin. Poly(IM-ac) whose imidazole content is less than 60 mol% and neutral imidazole

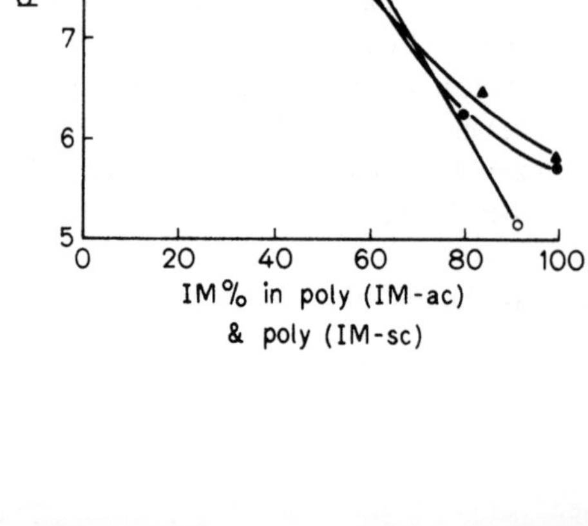

Fig. 5. pK_1 values of poly(IM-ac) and poly(IM-sc)
- ● poly(IM-ac) in 28.5% ethanol aq. soln. (30)
- ○ poly(IM-ac) in water (32)
- ▲ poly(IM-sc) in 28.5% ethanol aq. soln. (31)

content is less than 80% has higher catalytic activity than poly4(5)-vinylimidazole), in both hydrolyses of the neutral substrate, PNPA(5), and the positively charged substrate, ANTI(6) (*32*). This suggests the existence of the IM-COO^- cooperation.

This consideration of the cooperation is supported by some evidence. The pK_1 value of imidazolyl moiety in the polymer increases with the content of carboxylate moiety in the polymer. Existence of a counter cation shields the carboxylate anion and it decreases the pK_1 value of the imidazolyl moiety in the polymer by NMR study (*79*). Results are shown in Figs. 5 and 6.

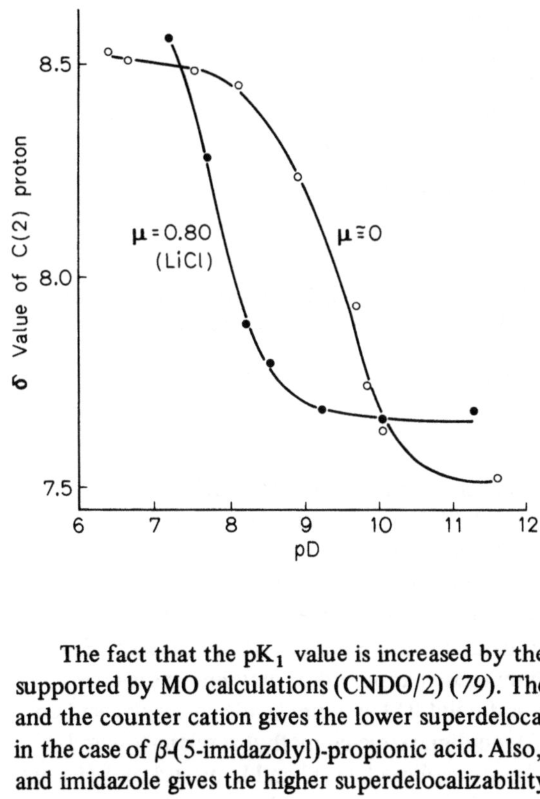

Fig. 6. Plots of values of C(2)proton of imidazole moiety in poly(IM-ac) (ac content = 81 mol%) against *pD*. (0.63 Meq. for imidazole moiety; room temp.)

The fact that the pK_1 value is increased by the COO^- anion in the polymer is supported by MO calculations (CNDO/2) (*79*). The closer distance between COO^- and the counter cation gives the lower superdelocalizability value of N(3) in imidazole, in the case of *β*-(5-imidazolyl)-propionic acid. Also, the closer distance between COO^- and imidazole gives the higher superdelocalizability value of N(3) in the imidazole, being adopted as a model case of 4-methylimidazole and the acetate anion system. Results are shown in Fig. 7.

In general, the activation of the nucleophile in the catalysis is promoted by the hydrogen bond formation with a general base. As is stated above, poly(1-methyl-5-vinylimidazole), poly(1M-5IM), has no hydrogen to form the hydrogen bond, so that the activation is not observed in this polymer. The hydrogen bond formation would be formed with a water molecule which is laid out between two functional groups. Accordingly, a contribution of a water molecule could be considered in the IM-COO^- cooperation. The contribution of a water molecule does not lead to a different conclusion in the calculation.

Molar extinction coefficients of both protonated and neutral imidazole moietis increase with the content of carboxyl groups in the poly(4(5)-vinylimidazole-co-

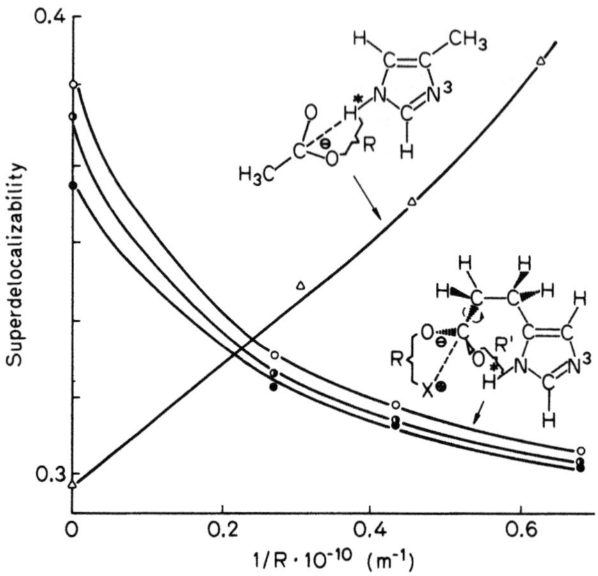

Fig. 7. Plots of superdelocalizabilities of N(3)'s in β-(5-imidazolyl)-propionic acid affected by the counter ion and in 4-methylimidazole affected by the acetate as the function of the distances.
○ gauche (R' = 1.828x10^{-10}m); ◐, gauche (R' = 2.622x10^{-10}m); ●, trans(R' > 3x10^{-10}m)

acrylic acid), poly(IM-ac) (*32*). This also suggests the cooperation IM-COO$^-$.

pK_1 value of imidazole moiety in poly(4(5)-vinylimidazole-co-vinylsulfonic acid), poly(IM-sc), also increases with the content of the sulfonic acid group in the polymer, as is shown in Fig. 5. However, the cooperation in the catalysis is not observed in the hydrolysis by the poly(IM-sc). The polymer is highly alternative, and the inactiveness of the isolated neutral imidazole moiety was considered (*31*).

IM-COOH-OH cooperation. Polymers such as poly(4(5)-vinylimidazole-co-γ-vinyl-γ-butyrolactone), poly(IM-la), and poly(4(5)-vinylimidazole-co-acrylic acid-co-vinyl alcohol) derived from poly(4(5)-vinylimidazole-co-methyl acrylate-co-vinyl acetate), both of which contain imidazole, carboxylic acid and hydroxyl moieties are synthesized and studied as a model of α-chymotrypsin (*29*). The former has a relatively ordered sequence and the latter has a random one. Results are tabulated in Table 11. The polymers cited in the Tabel contain a similarly low quantity of imidazole moiety, so that the cooperation of two subsequent imidazole moieties need not be discussed. Polymers such as L-84, L-68, M-83 and A-84 have higher catalytic activities than the polymer V-82. This suggests that the catalytic activity of the imidazole moiety in the polymers is much promoted by the carboxylate moiety in the polymers. The catalytic activities of L-84 and L-68 which have an ordered sequence are more than twice as high as that of M-83, having a random sequence. From these results it is concluded that the introduction of the hydroxyl moiety which has little cooperative effect on the imidazole moiety in V-82 in this reaction condition into imidazole and carboxylate – containing polymer, increases

Table 11. Comparison of the catalytic activities of polymers containing imidazole, carboxylate, hydroxyl moieties or two of them

Polymer	Composition (mol%)			k ($M^{-1} \cdot min^{-1}$)	
	IM	COOH	OH	For PNPA	For ANTI
L-84	0.086	0.457	0.457	13.7	5.9
L-68	0.19	0.405	0.405	13.0	5.5
M-83	0.17	0.42	0.41	6.0	2.9
A-84	0.165	0.835	0	3.3	2.4
V-82	0.18	0	0.82	1.5	0.1

20% ethanol; pH = 8.0; μ = 0.5; 30 °C;
Polymers

L

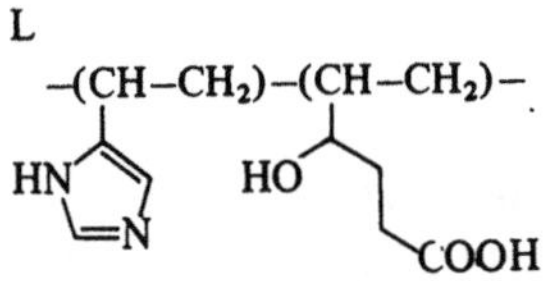

M

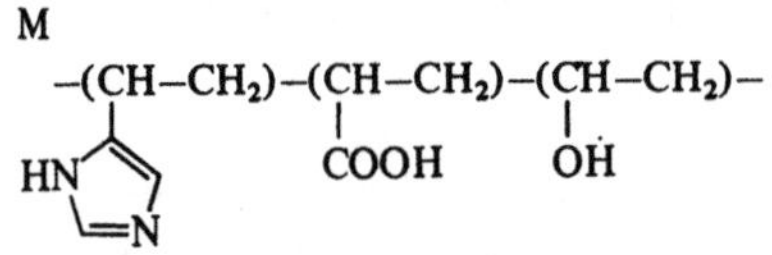

A

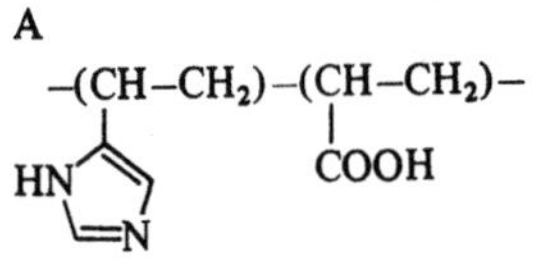

V

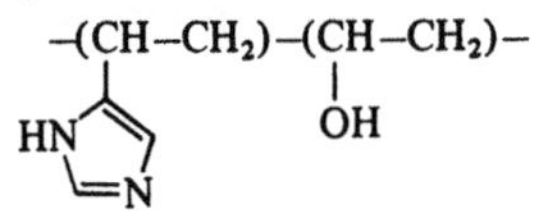

its catalytic activity. Also, the significance of the ordered sequence of IM-COOH-OH for the catalysis is suggested.

PY-PY cooperation. In the study of the hydrolysis of the neutral substrate, 2,4-dinitrophenyl acetate, by poly(4-vinylpyridine), the hydrolysis rate profile shows an upward curvature as the fraction of the neutral pyridine residue increases (*41*). It is considered that an increase in the nucleophilicity of the pyridine moiety is caused by the decrease of charge density in the polymer.

Im in His – amino acid residue cooperation. Though the catalytic activity of histidine residue, Im in His, is lower than that of imidazole, a cooperation between the Im in His and the amino acid residue in polymeric molecule would exist. Catalytic activities His-containing amino acid dimers in the hydrolyses of PNPA(5) give evidence of the cooperation. That is, the catalytic activity per His residue is increased by the other amino acid residue, in the order of magnitude;

Tyr > Lys > β-Ala > Cys > Asp > Glu > Ser (*20*).

The maximum increasing rate is 3.18 and the effect of polymerization is not remarkable. Catalytic activities His-containing polypeptides are also studied (*80*). As is shown in Table 12, a cooperation can be seen.

Compared with the activity of poly(L-His-co-L-Asp-co-L-Ser), the activity of L-Tyr-L-Ala-L-Ser-L-His-L-Asp is relatively high and reaches $92 M^{-1} \cdot min^{-1}$ (*81*).

Table 12. Catalytic activities of His-containing polypeptides for the hydrolysis of PNPA

Polypeptide	pK_1	Neutral Im in His (%)	k/neut. Im ($M^{-1} \cdot min^{-1}$)
(Imidazole)	7.2	77	45.9
(L-His · HCl)	6.0	98	5.8
Poly(L-His) (low mol.wt.)	5.6	99	9.6
Poly(L-His) (high mol.wt.)	5.9	98	5.6
Poly(L-His-co-L-Glu)	7.0	84	6.1
Poly(L-His-co-L-Ser)	6.4	95	10.2
Poly(L-His-co-L-Asp)	7.0	84	8.9
Poly(L-His-co-L-Asp-co-L-Ser)	6.7	91	3.7
Poly(L-His-co-L-Glu-co-L-Lys-co-L-Ser)	6.2	97	3.7

pH = 7.73; 0.2M phosphate buffer; 3.22% dioxane

In a comparison of catalytic activities of L-Ser-L-His-L-Asp and L-Ser-(γ-aminobutyl)-L-His-(γ-aminobutyl)-L-Asp, the activity of the latter is thrice as high as the former: The latter has the activity of $k = 147\ M^{-1} \cdot min^{-1}$ and the former has that of $k = 45\ M^{-1} \cdot min^{-1}$ (*18*). Though pK_1 values of these peptides are unknown, it is suggested that the flexibility of the peptide chain affects the cooperation.

Molecular Relay System

A clear-cut molecular relay system is materialized with the combination of two nucleophile moieties such as hydroxamic acid and imidazole introduced in a polymer (*38, 40, 82*). Hydroxamate anion has a high nucleophilicity but deacylation of the acylhydroxamate is very slow. The significance of the deacylation step in the hydrolysis in polymer catalyst is pointed out in several works, for example, the deacylation of acetylated poly(4(5)-vinylimidazole) (*83, 84*) and the aminolysis of nitrophenyl ester of poly(styrene-co-acrylic acid) (*85–87*). To promote the deacylation velosity of the acylated hydroxamic acid imidazole moiety is introduced in the hydroxamic acid-containing polymer. The polymers are shown in the following page.

Three active species are also identified (*82*). That is, the neutral imidazole, the neutral hydroxamic acid cooperated with imidazole, and the hydroxamate anion, which are illustrated in the following page with their catalytic activities, k./

The neutral hydroxamic acid does not show a catalytic activity as a nucleophile, so that the activation of the hydroxamic acid by imidazole is remarkable in these polymers.

In addition, as is shown in Table 13, the deacylation which is represented in k_d is increased by a factor of more than 10^3 in poly(PHA-IM-am), compared with poly(PHA-am). As is observed in the deacylation which is catalysed by imidazole moiety in the case of α-chymotrypsin, the catalysis by imidazole moiety in the deacylation of the acylated hydroxamic acid is considered. The turnover number, $k_{turnover}$, in poly(PHA-IM-am) is also high. The deacylation rate of poly(PHA-IM-am)

poly (MHA–IM)

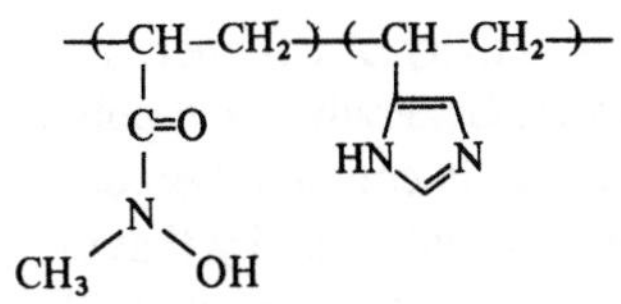

poly (PHA–2MIM–am)

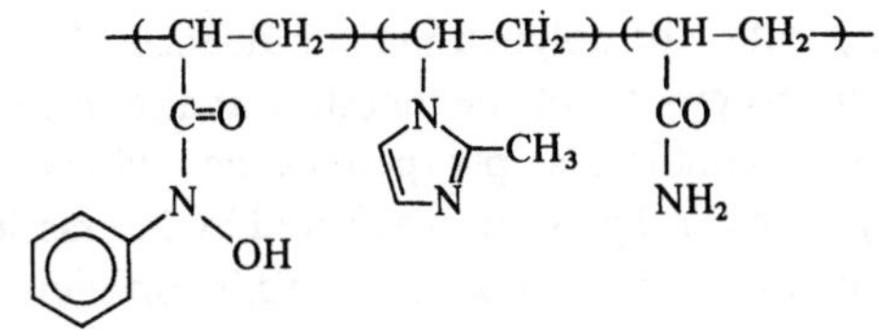

poly (PHA–IM–am)

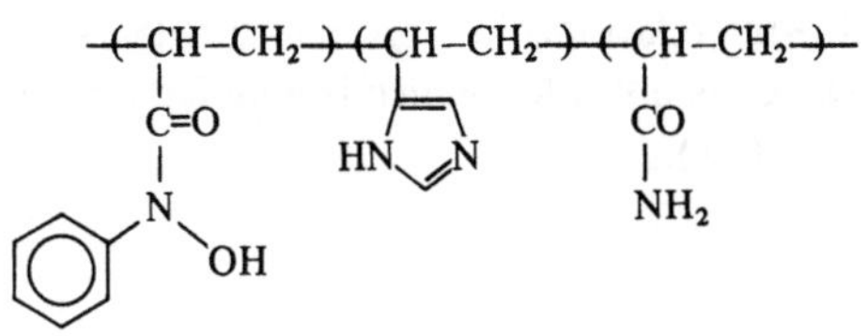

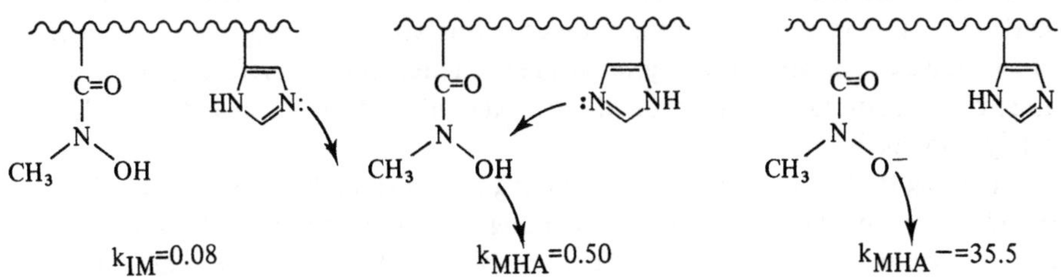

(k denotes the rate constant for PNPA(5) in $M^{-1} \cdot sec^{-1}$)

Table 13. Hydrolyses of PNPA by hydroxamic acid-containing polymers

Polymer	pH	k_a ($M^{-1} \cdot sec^{-1}$)	k_d (sec^{-1})	$k_{turnover}$ (sec^{-1})
Poly(PHA-IM-am) 5:35:60	8.04	7.9×10^{-1}	9.6×10^{-3}	7.8×10^{-5}
	9.01	2.59	2.64×10^{-2}	2.6×10^{-4}
Poly(PHA-am) 10:90	8.05	1.02	5.6×10^{-6}	5.3×10^{-6}
Poly(IM-am) 10:90	8.09	8×10^{-2}	10×10^{-4}	8×10^{-6}

increases with pH, and it allows a comparison with the deacylation rate in the case of α-chymotrypsin by pH = 9. However, whether this step proceeds in the nucleophilic or in the general base catalysis has not yet been concluded.

A similar deacylation mechanism is proposed in a micellar catalysis. Lauryl-imidazolylhydroxamic acid in a cationic micelle gives a high acylation rate, deacylation rate and turnover number, which are estimated to be 159 $M^{-1} \cdot sec^{-1}$, 4×10^{-3} sec^{-1} and 3.2×10^{-3} sec^{-1}, respectively (*88*).

3.3. Effect of Polymer Configuration

Configuration of polymer much influences the catalytic activity of polymer, not only in the process of the binding but also in the catalysis. Generally, vinyl polymers are more flexible than polypeptide, accordingly, the change in configuration occurs easily. The configuration is affected by interactions of side chains and by that of the polymer and the substrate. This is considered in relation to the concept of the induced fit (*90*). And then it is necessary to account for the dynamic configurational change in the catalysis.

For polyelectrolyte, in general, the following relationship is given in the degree of ionization, α, and the apparent ionization constant, pK_a, which is popularly called the modified Henderson-Hasselbach equation (*51*).

$$pK_a = pH + n' \log [1-\alpha)/\alpha]$$

where the deviation of n' from unity is a measure of the magnitude of the electrostatic effects, that is, relates to the excess electrostatic free energy for the removal of an equivalent of protons at a given degree of ionization. Therefore, using n' value as a measure, considerations on the dynamical configuration change of the polymer catalyst are made.

Turning to the description of the binding by hydrophobic interaction in 3.1. (p. 63), vinylpyrrolidone polymers containing benzimidazole [poly(ABI-pr)] and phenylimidazole [poly(API-pr)] have n' value of almost unity; on the other hand, acrylamide polymers containing benzimidazole [poly(ABI-am)] and phenylimidazole [poly(API-am)] have n' value of above unity. Intrinsic viscosities of the former are not influenced by the coexisting of urea which has a tendency of hydrophobic interaction cleavage, and those of the latter are increased by urea (*35, 37*). Therefore, in the latter, it is suggested that an aggregation occurs among large hydrophobic moieties, as is shown in III in Fig. 3. The extent of the aggregation increases with the content of the hydrophobic moiety also. That is, the aggregation occurs between the moieties far apart along the polymer chain and gives relatively large viscosity change. In the former the hydrophobic interaction between imidazole and pyrrolidone moieties are considered, as is shown in II in Fig. 3. However, poly(1-vinyl-2-methylimidazole-co-N-vinylpyrrolidone), poly(2MIM-pr), show a large conformational change in the binding of the substrate, as is shown in I in Fig. 3. In the case of II, a large binding constant, K, and a low catalysis rate, k_{cat}, are shown. On the other hand, a relatively low binding constant, K, and a relatively high catalysis rate, k_{cat}, are shown, in the case of III. This is explained as showing that the binding is not so efficient, since an intramolecular interaction between hydrophobic moieties takes place. Relatively less change in k_{cat} with change in K implies that the hydrophobic binding does not much influence the rate of the catalysis.

Poly(4(5)-vinylimidazole) shows a dramatical configuration change by the solvent system and a remarkable change in catalytic activity (*24, 56*). The apparent pK_1 value for the polymer decreases by the addition of ethanol up to 50 vol% and a minimum value of pK_1 occurs in the region of 50 vol% ethanol-water, and the addition of ethanol to water leads to an increase of the pK_1 value; whereas, pK_1 value of imidazole decreases monotonously with ethanol content. A drastic change in n′ value in the titration in high ethanol content is also observed. Intrinsic viscosity of the polymer increases with ethanol content and shows a maximum, then decreases with the ethanol content. These phenomena lead to the consideration that the configuration of the polymer changes drastically with the content of ethanol; a shrinkage of the polymer in solvent of low ethanol content might be achieved by intra- and intermolecular nonpolar interactions leading to the exposure of imidazole moiety, whereas its shrinkage in a solvent of high ethanol content is attributed primarily to intra- and intermolecular interaction between IM^+ and IM moieties leading to the burying of imidazole moiety, as is illustrated in Fig. 8.

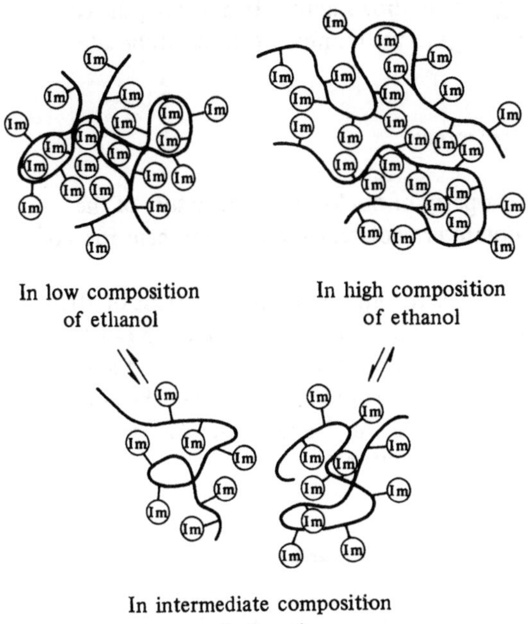

Fig. 8. Schematic illustration of the configuration of poly(4(5)-vinylimidazole) in ethanol-water. From J. Amer. Chem. Soc., *93*, 3226 (1971)

In this connection, both kinetic schemes, such as the Michaelis-Menten kinetics in the solution of which ethanol content is below 40% and the second-order kinetics in the solution of which ethanol content is above 40%, are observed. In the hydrolysis of the negatively charged substrate, NABA(3), the rate of the acylation in the "shrunk-formed" poly(4(5)-vinylimidazole) is a thousand times higher than that in the "loose-formed" polymer. This suggests that the catalysis is much influenced by the polymer configuration.

The distance of the catalytic point from the polymer chain is very important. Hydrolyses of PNPA(5) by poly(l-methyl-4-vinylimidazole), poly(lM-4IM), and poly(1-methyl-5-vinylimidazole), poly(1M-5IM), are carried out (*54*). Poly(1M-5IM) shows a certain catalytic activity, while poly(1M-4IM) does not show the catalytic activity. Also, by NMR study, that the methyl group in poly(1M-4IM) is located apart from the polymer chain and the methyl group in poly(1M-5IM) is located close to the chain is discussed. This result indicates the significance of microstructure of the catalytic site.

The significance of the polymer configuration is also pointed out in a partially protonated poly(4-vinylpyridine) (*64*). Of partially benzylated poly(4-vinylpyridine) the reduced viscosity of the polymer shows a sharp minimum at a definite degree of benzylation, and under this condition all of the free pyridine moiety acts as the catalyst for the hydrolysis. The polymer, having the degree of polymerization of 2400 shows the minimum reduced viscosity with the benzylation being at about 86%, and the highest catalytic activity. There is some discussion by the same authors (*90*) of the catalytic species which might be hydroxylamine. In any case, the conformation of the polymer causes the change in the number of the catalytic site.

3.4. Effect of Environment of Polymer Domain

Polymer environment influences the property of nucleophile, the nature of the catalyst and the stability of the reaction intermediate. Here the significant roles of polymers are considered.

Table 14 shows activation parameters of the intramolecular reactions of Michaelis complexes of which the hydrolyses of ABA(7) by poly(ABI-py) and poly(ABI-am) are the same systems as shown in Table 6 (*35*). Both the polymers and the low molecular weight analogues have the hydrophobic binding properties. The polymer catalysts show smaller $\Delta H^{\ddagger}$ values by $1 \sim 3$ kcal · M^{-1} and greater $\Delta S^{\ddagger}$ values by $6 \sim 10$ eu, than those of the small molecule catalysts.

Table 14. Thermodynamic parameters of the intramolecular reactions of Michaelis complexes

Polymer	$\Delta H^{\ddagger}$ (kcal · M^{-1}	$\Delta S^{\ddagger}$ (cal · M^{-1} · T^{-1})	$\Delta G^{\ddagger}$ (300 °K) (kcal · M^{-1})
Poly(ABI-py) (Im = 1.94%)	5.26	−58.0	22.83
Poly(ABI-am) (Im = 8.91%)	4.17	−60.0	22.35
Low mol. # analog.			
(I)	7.31	−50.4	22.55
(II)	6.26	−52.0	21.96

(I) and (II) are shown in Table 6.

It is found that the rate of the nucleophilic attack of imidazole, k is related to its acid dissociation constant, pK_1, by the relationship

$$\log k = 0.8\ pK_1 - 4.30$$

in the hydrolysis of PNPA(5) (*91*). To obtain the corrected intrinsic nucleophilicity of imidazole moiety in the polymers, the following equation is derived using the above relationship (*35*)

$$\Delta G^{\ddagger}_{corr} = \Delta G^{\ddagger} + 0.8\,(pK_1 - 5.60) \times 2.30\,3\,RT$$

For the hydrophobic binding group-containing polymers such as poly(ABI-pr), poly(ABI-am) and poly(API-am-ap), poly(API-pr), poly(API-am) and poly(API-am-ap), $\Delta G^{\ddagger}_{corr}$ and ΔG_u which are obtained in the binding are correlated fairly well, whereas, plots for the small molecules deviate from the correlation. The relationship between $\Delta G^{\ddagger}_{corr}$ and ΔG_u for the polymer catalysts is given by

$$\Delta G^{\ddagger}_{corr} = -0.82\,\Delta G_u + 18.9$$

It was indicated that $\Delta G^{\ddagger}_{corr}$ and ΔG_u are inversely correlated and that the transition state of the intracomplex process is stabilized as the extent of the hydrophobic binding increases and it was concluded that the catalytic property of the polymers is largely affected by the hydrophobicity of the catalytic site, apart from the nucleophilicity of imidazole moiety in those polymers (*35*). Also, the large entropy of activation was considered as showing that the intracomplex formation takes place overcoming an entropic barrier by a breakdown of the structural water around the catalytic point.

In general, the environment of polymer domain influences activities of nucleophile, electrophile and general base catalysis. It is easily understood that unusual pK values of amino acid residues in enzymes are given, as is shown in Table 15 (*92*).

Table 15. pK values of amino acid residues in several enzymes

		pK
Lysozyme	Asp (52)	3.5
	Glu(35)	6.3
Chymotripsin	His(57)	6.7
Subtilisin	His(64)	7.2
Carboxypeptidase A	His(196)	7.7
	His (96)	9.1
	Glu(270)	7.2
Ribonuclease	His(12)	6.2
	His(119)	5.8

Such shifts of pK values are considered in terms of the following causes. A coulombic field which is formed by charged moieties on the polymer, polarizes the catalytic moiety and changes its proton affinity. The pK value of the nucleophile would increase in an anionic field and decrease in a cationic field. The ionization of the nucleophile is promoted by base. The hydrogen bond formation in the nucleophile influences its pK value directly as the charge relay system. A hydrophobic field also influences the pK value of the nucleophile. In this case, the stabilization of the charged or polarized state is so delicate that the direction of shift is not restricted to one direction. Brönsted law shows us that a general acid having a lower pK value and a general base having a higher pK value have higher catalytic activities. As the shift of pK depends on a polarization in the hydrophobic moieties, the hydrophobic environment gives a preferable shift in pK to acid having a lower pK and base having a higher pK. Accordingly, these acids and bases can play the roles of the catalyst in a moderate pH condition. Given the difference between the ionized and the neutral states in free energy, $\Delta G = G_{ionized} - G_{neut} > 0$, then the pK values of the acid and the base are shown as $\Delta G/RT$ and $14 - \Delta G/RT$, respectively. In the hydrophobic environment, since the stabilizability of the ionized state induced by the polarization of solvent is less, the ΔG value increases, compared with that in water, so that the pK value of the acid increases and the pK value of the base decreases. Accordingly, the number of the active catalytic moiety of the neutral state increases which in turn promotes the catalytic activity.

The significance of the coulombic field is observed (*79*). Figure 9 shows the catalytic activities of poly(4(5)-vinylimidazole-co-acrylic acid), poly(IM-ac), and low molecular weight imidazole derivatives in hydrolyses of various substituted phenyl acetates of which leaving groups have different activities.

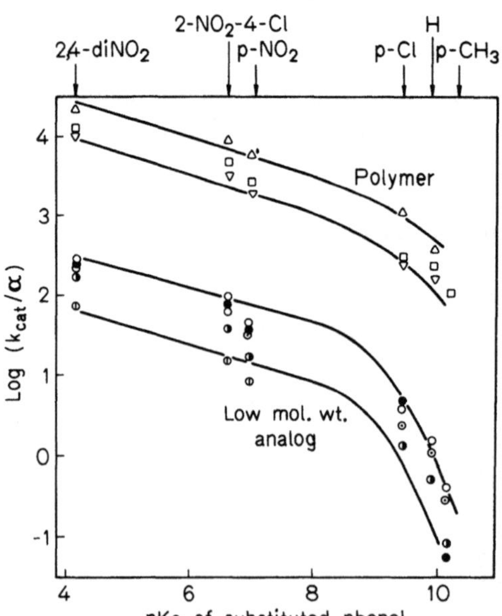

Fig. 9. Rates of hydrolyses of substituted phenyl acetates by various catalysts against pKa's of leaving phenols. [Cat] = 4×10^{-3}M; [Ester] = 2×10^{-2}M ●, imidazole; ○, β-(5-imidazolyl)-propionic acid; ⊙, α-hydroxy-β-(5-imidazolyl)-propionic acid; ◑, *cis*-urocanic acid; ⦶, 4(5)-vinylimidazole; Δ, poly(IM-ac) (ac = 81 mol%); □, poly(IM-ac) (ac = 74 mol%); ▽, poly(IM-ac) (ac = 65 mol%).

The rate constant in the hydrolysis by every catalyst decreases with increase of basicity of the leaving group of the substrate, ArO^-. And it shows a precipitous fall of the rate when the basicity of ArO^- is stronger than a certain value. The figure is classified roughly into polymer and low molecular weight groups. The precipitous fall of the rate suggests a change in the transition state that is expected for the rate-determining process. The fact that the attack of imidazole onto the acetate is the rate-determining step with a good leaving group and that the breakdown of an addition intermediate is the rate-determining step with a poor leaving group, is shown in the investigation of imidazole catalyzed and alkaline hydrolyses of a series of acetates (*93*). In the present case, hydrolysis of the ester of which the pK_a value of the leaving group is high, *i.e.* a breakdown of the addition intermediate is rate-determining, slopes of the curves in the polymer catalysts are lower than slopes in the low molecular weight catalysts. This is presumably owing to higher pK_1 values of polymer catalysts rather than to low molecular weight catalysts. In addition, the Brönsted plots of the hydrolyses, that is the logarithm plot of the rate constant against the pK_1 value of the nucleophile, shows linear relationships in each catalyst. The fact that hydrolyses of all esters by a catalyst lies on the same line suggests that the nucleophilicity of the nucleophile affects both the breakdown of the addition intermediate and the nucleophilic attack. The slope of the line for the polymer catalyst is much larger than that for the low molecular weight catalysts; the former is about 4 and the latter is about 1. Moreover, the catalytic activities of the polymers are 100 times higher than that of the low molecular weight catalysts. These results suggest that the catalytic activity of the polymer catalyst is caused by the nucleophilicity of the imidazole moiety and an additional positive effect which cannot be found in the low molecular weight catalyst. That is to say, the addition intermediate which is thought to be more polar than the neutral substrate is stabilized in the high electrostatic field and high ionic strength in the polymer environment, and the leaving of ArO^- from the addition intermediate is facilitated by the electrostatic repulsion between the leaving ArO^- and the carboxylate anions in the polymer. On the other hand, n' value (the coefficient in the modified Henderson-Hasselbach equation) for the carboxylate moiety in the polymer becomes larger with the content of the carboxyl moiety where the deviation of n' value from unity is the measure of the magnitude of the electrostatic effect, so that the strong anionic field that is owed to carboxylate anion exists in the polymer domain. It is expected that these factors contribute to the higher catalytic activity and the increase in the slope in the Brönsted plot for the polymer.

Stated above, the pK_1 values of the imidazole moieties in the poly(IM-ac) are much larger than that of the low molecular weight derivatives. This is also confirmed by NMR measurement and theoretical calculations (*79*). Plots of chemical shifts of C(2) proton of the imidazole moiety against pD show a lower pK_1 value of the imidazole moiety when a considerable amount of salt is added. It might be considered, as the addition of salt reduces the strong anionic field in the polymer domain, that the decrease of the pK value is caused by the weakening of the anionicity of the field. Theoretically, the proton affinity and the superdelocalizability are calculated, adopting model systems such as 2-(5-imidazolyl)-propionic acid – counterion and 5-methylimidazole – acetate ion systems. The superdelocalizabilities of N(3) of the imidazole moieties in both those molecules increase with a withdrawing of the counterion from the carboxylate and with an approaching of the carboxylate ion, as is described above (see Fig. 7). These results also lead to the conclusion that the stronger anionic environment of the polymer gives the higher pK value of the imidazole moiety in the polymer.

An appreciable change in the pK value of the indicator having an opposite charge to that of the polyion is observed (*170*). The change is dependent on the property of the polyion and the interaction of the indicator with the polyion.

Myristoyl-methylhydroxamic acid(VIII) which has an anionic catalytic moiety shows an extremely high nucleophilicity in partially alkylated poly(4-vinylpyridine) (IX). The rate of the acylation in the hydrolysis of PNPA(5) is given in k_a = 100 $M^{-1} sec^{-1}$ (pH = 8, 30 °C) (*94*). This suggests that the polymer environment, a cationic field, influences the anionic catalyst.

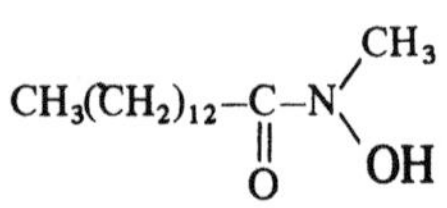

(VIII)

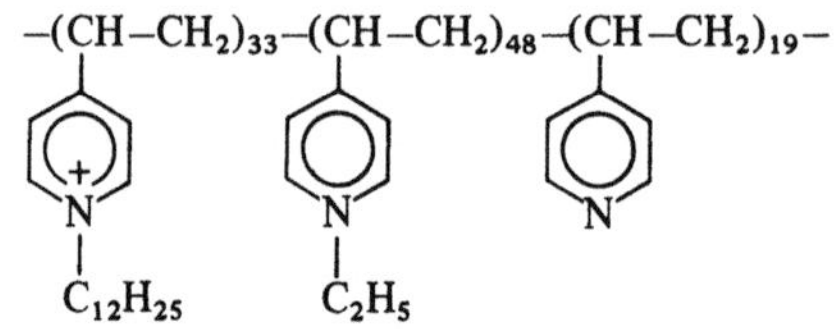

(IX)

Usually oxime has a high nucleophilicity in dissociated form, but it is useless in a moderate aqueous solution for its high pK value (pK = 11 ~ 12). The oxime moiety which is bound to poly(4-vinylpyridine) by the quarternization shows a lower pK value (pK = 8.5 ± 0.3) and high nucleophilicity in the hydrolysis of PNPA(5) (*95*). The catalytic activity of the polymer(X) reaches a half of that of α-chymotrypsin showing k_a = 200 $M^{-1} sec^{-1}$ at pH = 8, where m is 12.

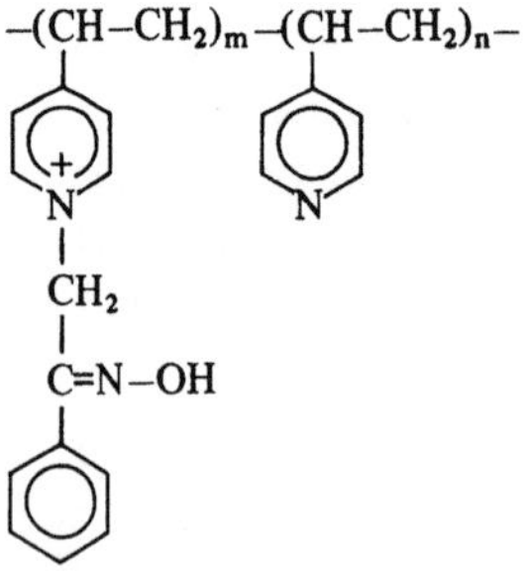

(X)

Of hydroxamic acid – containing polymers such as (XI) and (XII), decreases in pK value are observed (*94*).

Similar polymer environment effects can be seen in micelle catalyst systems which are an oxime (*96*), a thiol (*98*), a hydroxamic acid (*99–101*) in cationic micelles and imidazole-containing cation micelle (*97*). In these catalyst systems hydrolyses of PNPA(*5*) are extremely increased. Decreases of pK_a values and dehydration at the catalytic moiety are suggested.

The polymer chain has a hydrophobic environment itself. The environment makes the pK_1 value of a nucleophile-containing polymer decrease most, as is shown in Table 16.

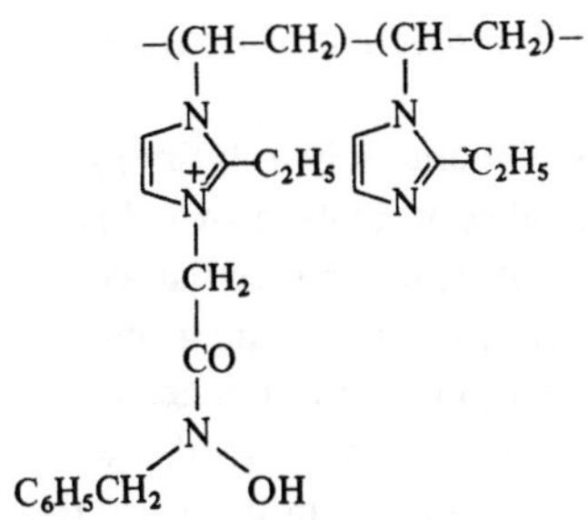

(XI)

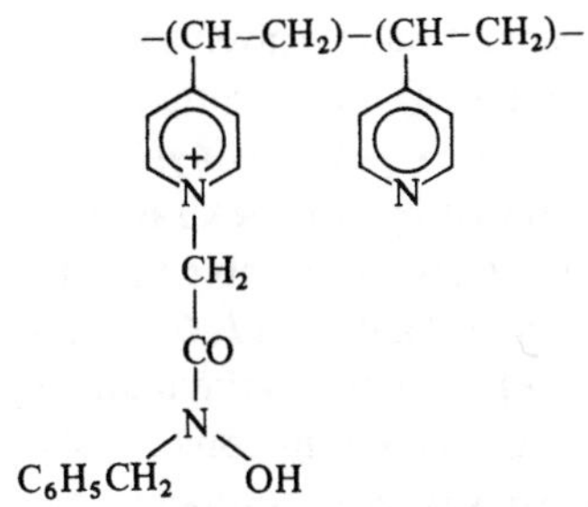

(XII)

Table 16. pK_1 values of polymers and their low molecular weight analogues

Polymer	pK_1	Low Mol. Wt. Analog.	pK_1
Poly(4(5)-vinylimidazole)	6.2	5-Methylimidazole	7.45
Poly(N-vinylimidazole)	4.16	N-Ethylimidazole	7.2
Poly(NIM-pr) (1:25)	5.75		
Poly(NIM-pr) (1:13)	5.63		
Poly(4-vinylpyridine)	4.33	4-Ethylpyridine	6.22

4. The Cooperative Interaction Between Metal and Nucleophile in Nucleophile-Containing Polymers

Nucleophile-containing polymer complexes have peculiar properties in addition to the properties described in 3. Complexizations under restraint of polymeric chain would give different properties from those by low molecular weight legands. Generally, the composition and the stabilization of the complex of the polymer differ widely from those of the low molecular weight, nevertheless, legand and metal are the same. The peculiar catalytic activity accompanying the steric effect would be given. In addition, the polymer holds the metal without coagulation.

4.1. Coordination

Poly(acrylhydroxamic acid)-Cu(II) complex has a high catalytic activity in the decomposition reaction of hydrogen peroxide. From the comparison of complex formation constants (stabilization constants) both in the polymer and the low molecular weight molecule, it was considered that the high activity is owed to a partly dissociated or a solvent coordinated moiety (*104, 105*). Poly(methacryloylacetone)-Cu(II) complex is also investigated (*108*).

The significance of axial coordination sites of Cu(II) in poly(L-histidine)-Cu(II) complex for oxidation and reduction reaction is pointed out (*106, 107*). A loosely axial coordinated Cu(II) ion gives a high activity.

A similar consideration was given to the decomposition reaction of hydrogen peroxide by poly(L-lysine)-Cu(II) complex (*109*).

Poly(4-vinylpyridine), poly(N-vinyl-2-methylimidazole) and poly(L-lysine) give polymeric hemochromes by complexizations with an axial coordination site of Hem. The polymeric complex is more stable than the complex of its low molecular weight analogue, as is shown in Table 17 (*110*). Penta-coordination structures are given in poly(4-vinylpyridine) and poly(N-vinyl-2-methylimidazole), in addition, a hexa-coordination structure is given in poly(L-lysine). It is considered that one of the axial coordination sites is vacant or loosely coordinated to a solvent molecule in the case of poly(4-vinylpyridine) and poly(N-vinyl-2-methylimidazole). The coordination number of the axial-base in poly(L-lysine) being 2 was considered as a high pK_a value of the polymeric legand or a regular polymer conformation in the solution. The addition of poly(ethylene glycol) to poly(L-lysine) gives a polymer complex and its coordination number to Hem is reduced to 1.

In the polymerization of acrylonitrile by the imidazole-Cu(II) complex, $Cu(IM)_4X_2$ ($X = ClO_4^-, NO_3^-, Cl^-, Br^-$), the complexes which have a relatively weak coordination with NO_3^- and ClO_4^- have considerable catalytic activities; however, the complexes having strong coordinates such as Cl^- and Br^- have no catalytic activities. In this study it is also suggested that the polymerization proceeds by a coordination of acrylonitrile at the axial site (*110*).

As is described above, the property of the axial coordination site influences the catalytic activity in the hexa-coordinate compounds.

A similar effect is observed in the hydrogenation of olefin by poly(styrene)-titanocene (reduced form) (*112*). A not-fully coordinated species is considered by a long-life ESR signal with respect to the catalytic activity in this case.

Generally, Wilkinson complexes which are coordinated with polymeric legands show a relatively high stability in reactions. This is explained as a contribution of the hydrophobic environment of the polymer (*113, 114*). Besides, the polymer plays a role of holding metal without its coagulation. Poly(naphtoquinone)-K-$FeCl_3$ in the ammonia synthesis (*115*) and poly(acrylic acid)-Rh(III) in the hydrogenation are shown as examples (*116, 117*).

In addition, polymeric phosphine-Pd(II) complexes are shown to be useful in the hydrogenation, hydrosilylation and hydroformylation of methyl ester of unsaturated fatty acid (*118*).

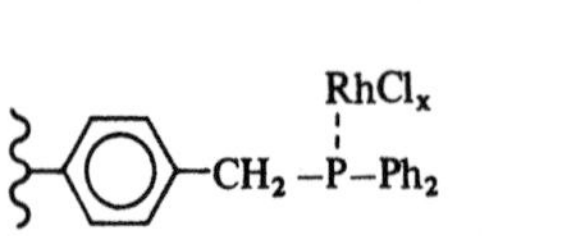

Hydrogenation, Hydrosilylation

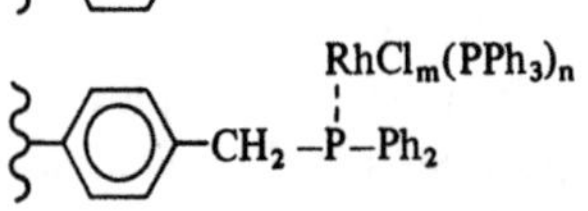

Hydrogenation

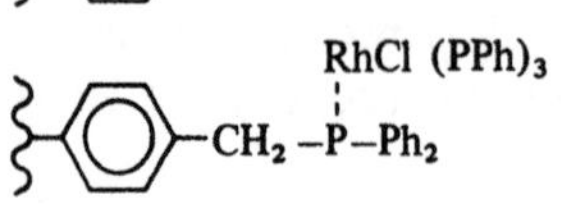

Hydrosilylation

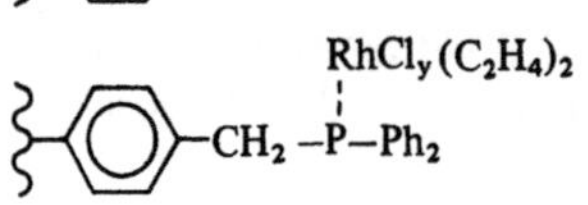

Hydrogenation, Hydrosilylation

Table 17. Equilibrium constant and coordination number of axial-base in complex formation of hemochrome

Axial base	Eq. const, K (M^{-1})		Coord. no. (n)		Solvent
	Fe(III)	Fe(II)	Fe(III)	Fe(II)	
Pyridine	6.04	5.13×10	1.02	1.11	DMF-MeOH (7:3)
Poly(4-vinylpyridine)	3.75×10^2	1.14×10^3	0.95	1.32	
Imidazole	9.52	4.55×10	1.14	1.20	
Poly(N-vinyl-2-methylimidazole)	1.78×10^2	2.55×10^3	1.07	1.15	DMF-MeOH (1:3)
Partially quaternized poly(4-vinylpyridine)	2.60×10^2	2.77×10^4	1.30	1.25	H_2O-DMF (9:1)
Poly(L-lysine)	4.22×10^4 [a]	3.49×10^9	2.00	2.20	H_2O(pH = 12)

[a] $K(M^{-2})$ at 20 °C, 25 °C.

4.2. Binding

The Binding by Coulombic Interaction

Poly(L-histidine)-Cu(II) complex catalyzes oxidation reactions of ascorbic acid, homogentisic acid and p-hydroquinone with molecular oxygen at the condition of pH = 4 ~ 6 (*106*). These substrates which are neutral or negatively charged follow Michaelis Menten kinetics at the reaction. The overall reaction rate catalyzed by the complex is higher than that by Cu(II), as is shown in Table 18. On the other hand, the rate in the case of the positively charged substrate, phenylene diamine, is very low and the rate is reduced with the ionic strength. These phenomena suggest the existence of a binding between imidazolium moiety in poly(L-histidine) and the substrates.

Table 18. Oxidation of ascorbic acid, homogentisic acid and p-hydroquinone by poly(L-lysine)-Cu(II) complex

Substrate	pH	K_m (M^{-1})	k_{cat} (min^{-1})	k_{obs} (min^{-1}) PLL-Cu(II)	Cu(II)
Ascorbic acid	4.3	4.3×10^{-5}	6.0	1.0	1.6×10^{-1}
Homogentisic acid	5.0	6.0×10^{-5}	8.8×10^{-1}	1.0×10^{-1}	5.8×10^{-4}
p-Hydroquinone	4.8	4.0×10^{-4}	1.37	1.9×10^{-1}	2.0×10^{-3}

In the polymerization of xylenol by partially quaternized poly(4-vinylpyridine)-Cu(II) complex, the binding between the positively charged polymer and the substrate, xylenol anion, is observed. The overall rate catalyzed by the polymer is five times higher than that catalyzed by the pyridine-Cu(II) complex (*119–121*).

In the oxidation reaction of the anionic reactant, $[Fe(II)\text{-}EDTA]^{-1}$ with poly(4-vinylpyridine)-Co(III)en_2 complex(XIII), the coulombic binding between the reactant and the polymer is considered (*122–124*).

The influence of polyelectrolytes on the electron transfer reaction are investigated closely (*125–130*).

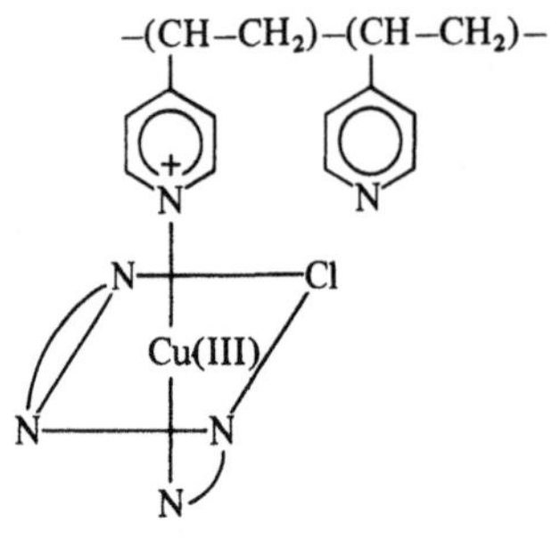

(XIII)

The Binding by hydrophobic Interaction

Using an inclusion ability of cyclodextrin, a metal-containing model catalyst as illustrated in Fig. 10 is investigated (*131*). Complexization of metal ion such as Ni(II) promotes the hydrolysis of PNPA(5) by a factor of about 1000.

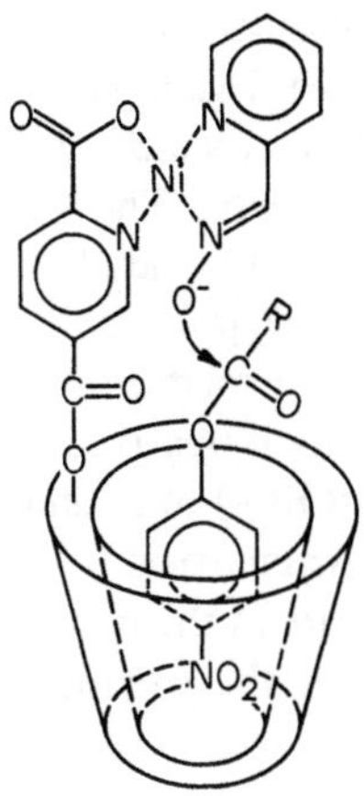

Fig. 10. Metal-containing cyclodextrin model catalyst

4.3. Steric Effect

In addition to the coordination and the binding the steric effect caused by the polymer legand influences the chemical reaction in metal-containing polymer catalysts; one is the specificity in the coordination and the other is the entropic effect in the active center. The former would control the catalytic activity of the active center.

Polymeric phosphin as described above (p. 90) gives *trans* structure and its catalytic activity is remarkable (*118*). It is suggested that a novel catalytic species which cannot be observed in low molecular complex is formed.

Polymeric Catalysts such as porphirin-containing poly-(L-phenylalanine)-Fe(III) complex in the dehydrogenation of p-phenylenediamine (*132*), poly(L-lysine)-protoporphirin Fe(III)-4,4′-dipyridyl complex in the oxidation of cytochrome (*133*), poly(L-lysine)-Cu(II) complex in the hydrolysis of oligophosphate (*134*), and poly(L-lysine)-Cu(II) complex in the oxidation of 3,4-dioxyphenylalanine and in the hydrolysis of phenylalanine (*135–137*) have their own structures.

Poly(ϵ-carbobenzoxy-L-lysine)-Cu(II) complex is used in the selective dehydrogenation reaction of alcohol (*138, 139*). X-Ray analysis shows that Cu(II) is coordinated with amide moiety in amino acid residue and peptide moiety in the helix chain, and gives the selectivity in the dehydrogenation (*139*).

Poly(L-lysine)-Cu(II) complex catalyzes the oxidation of D-3,4-dioxyphenylalanine (D-DOPA) much more than L-DOPA, with the condition that poly(L-lysine) gives α-helical configuration (*135, 136*). The α-helical structure serves to hold the cupric complex in a special configuration. The special asymmetrical structure leads the asymmetrical selectivity in the reaction.

Poly(L-lysine)-Cu(II) complex exhibits the asymmetrically selective catalysis on the hydrolysis of phenylalanine esters (*137*). In this case, the asymmetrical structure of the complex gives the asymmetrical selectivity in the reaction also.

In the hydrogenation of methyl acetoacetate by poly((s)-2-methylethyleneimine)-Ru(III) complex, an asymmetrical selectivity is observed (*140, 141*). On the other hand, complexes of poly((RS)-2-methylethyleneimine) and low molecular weight legands have no asymmetrical selectivities in the reaction. The polymer legand keeps its complex homogeneity during the reaction, besides, the low molecular weight legand detaches and separates the metal and the catalytic activity decreases. The polymer legand plays a role in the stabilization of the complex keeping the metal in a low valence state. The hydrogenation reaction is also achieved using poly(L-glutamic acid)-Ru(III) complex, and a remarkable result concerning the relationship between the conformation of the polymer complex and the selectivity is observed. The polymer complex which is a random structure gives the product R(−) form and that which is a helical structure gives the product S(+) form. The asymmetrical selectivity varies continuously with the content of the helical structure in the polymer complex. The selectivity is much influenced by the content of helix in the polymer having higher ratio of polymer and Ru(III).

In addition to this we have several examples of which the polymer conformation of the polymeric complex leads the asymmetrical selectivity: Hydrogenation reactions of l-methylcinnamic acid and l-acetamidocinnamic acid by several poly(L-amino acid)-Pd complexes are observed (*142–144*). Poly(L-valine) (β-form) and poly(β-benzyl-L-aspartate) (α-helix, sinistral) give dextrorotative products, and poly(L-leucine) and poly(β-benzyl-L-aspartate) (α-helix, dextral) do levo-rotatory products. Also, optical active poly-β-hydroxyl esters–Raney Ni catalyst (*145*) and Ion-exchange resin modified by optical active amino acid–metal complex (*146, 147*) are observed in asymmetrically selective hydrogenations.

As a hemoglobin model, using imidazole attached poly-(styrene) with the picket fence porphyrin-Co(II) (XIV) the reversible oxygen adsorption-desorption reaction is observed (*148, 149*).

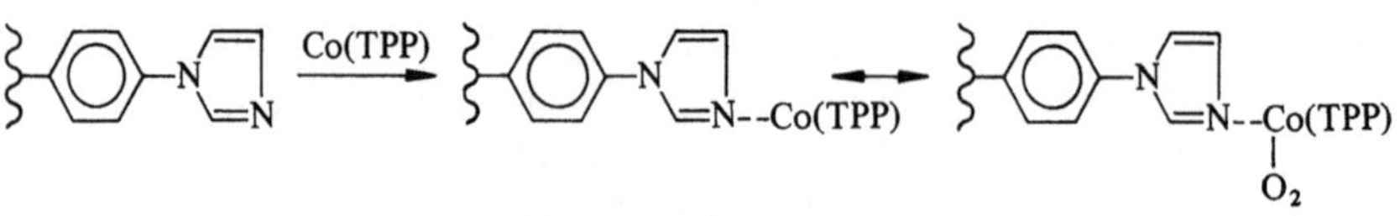

(XIV)

Polymer-bonded metalloporphirins are usually durable (*150, 151*). To avoid the interaction between pendant groups is one of the most important roles of polymers. In addition, it is reported that Fe(TPP) $(py)_2$ and Fe(TPP) $(pip)_2$–imidazole bound on silica gel complex are inhibited in the case of dimerization such as Fe(III)-O-Fe(III) formation (*152*).

Polymeric phosphin–Rh (XV) hydrogenizes several kinds of cyclic unsaturated hydrocarbon selectively, specifying the molecular size (*114, 153*).

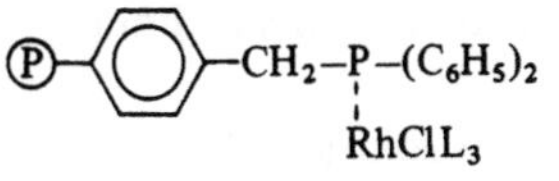

(XV)

Polymeric phosphin-Ni complex also has a selectivity in its catalytic activity (*154*). Such a sterically selectivity is shown in an intrapolymer electrophilic reaction of Fe-carbonyl-olefin complex (*155, 156*).

Copolymers (XVI) which have oxime moiety give complexes of Fe(II), Co(II) and Ni(II), and they adsorb and desorb molecular oxygen reversibly (*157*).

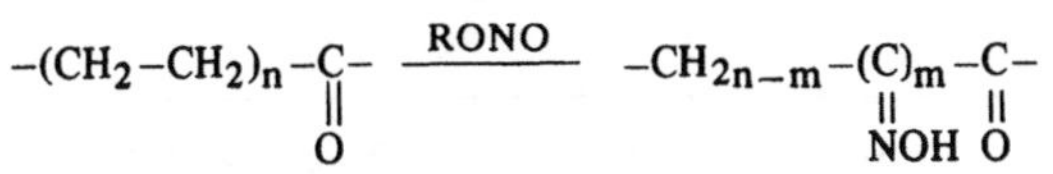

(XVI)

4.4. Legand-Metal Interaction

The polymeric legand suffers restriction by its polymer chain and differs from the monomeric legand. A noteworthy phenomenon transpires from the study of polymeric hemochrome (*158–160*). The physical property of the legand is important as well as the chemical property to give functionality. Magnetic susceptibility of the polymer–hemin (Fe(III)) complex shows an intermediate value of the high and the low spin states. The fact that a thermal equilibrium is set up between both spin states suggests the significant role of the polymeric legand. In the case of the monomeric legand such a phenomenon is not given. As is shown in Table 19, the energy difference between both spin states, ϵ, varies with the axial legand. State ratio (γ), enthalpy change and entropy change in the complex formation are also given in different values. Polyion complex which is composed of partially quaternized benzyl chloride (35%) and sodium poly(styrenesulfonic acid) coordinates hemin, its coordination number is the same as pyridine–hemin and poly(4-vinylpyridine)–hemin

Table 19. Various parameters of the transition of spin states for hemin complexes

Ligand	Ground state	ϵ (cm^{-1})	γ State ratio	$\Delta H^{\ddagger}$ (cal M^{-1})	$\Delta S^{\ddagger}$ (eu)
Cl	High	− 33	1.30	+ 95	−2.73
Pyridine	High	−111	0.98	+318	−2.15
PVP	High	−225	0.54	+649	−0.98
QPVP	Low	+ 32	0.29	− 92	+0.22
QPVP + NaSS	Low	+138	3.92	−398	−4.93
PLL[a]	Low	+ 93	0.0335	−268	+4.59

PVP, poly(4-vinylpyridine): QPVP, PVP partially quaternized by benzyl chloride (35%): NaSS, sodium poly(styrenesulfonic acid): PLL, poly(L-lysine).

[a] Hexa-coordination configuration.

complexes. However, electronic structures of iron ions are different. This is thought to be an influence of the polymer environment.

Generally, hem is readily oxidized by oxygen in an aqueous solution. The hydrophobic reaction in vivo takes place in a hydrophobic field which is formed by apo-enzyme. A stable hydrophobic field disrupts a solvation and makes the hemin stabilize. Poly(phenylalanine)–hemin shows higher catalytic activity than hemin in the oxidation of phenylenediamine (*132*). This is explained by the fact that the hydrophobic environment fomred by polypeptide plays a significant role in the stabilization of the hemin.

Table 20. Rate constant and activation parameters in oxygenation of pyridine and imidazole hemochromogens

Axial base	k (sec^{-1})	E_a (kcal M^{-1})	$\Delta H^{\ddagger}$ (kcal M^{-1})	$\Delta S^{\ddagger}$ (eu)	Oxygen saturation (%)
Pyridine	0.989×10^{-4}	11.0	10.4	−43.6	60
PVP	1.86×10^{-4}	26.9	26.4	+ 8.85	91
Imidazole	4.78×10^{-4}	10.1	9.44	−43.4	63
PMVI	7.76×10^{-4}	19.2	18.6	−13.2	93

PMVI, poly(N-vinyl-2-methylimidazole).
[heme] = 9.0×10^{-6}M: [Na_2SO_4] = 7.2×10^{-4}M: [Axial base] = 1.0×10^{-2}M:
Solvent, DMF/MeOH(9/1) + trace of water: 40 °C.

As is shown in Table 20, the rate of oxygen adsorption on the polymeric hem is higher than that of the monomeric hem (*158, 159*), and the total amount of the oxygen adsorbed on the polymeric hem is larger than that on the monomeric hem. Entropy values of both adsorptions are also different. These phenomena suggest the significance of the polymer environment. Oxygenation of heme is represented as follows,

$$Fe(II) + O_2 \overset{k_a}{\rightleftharpoons} [\overset{O=O}{\overset{|}{Fe}}(II) \cdot \overset{O-O^{\ominus}}{\overset{\oplus|}{Fe}}(III)] \xrightarrow{k_d} Fe(III) + O_2$$

The irreversible process (k_d) propagates with a weakening of the ionic bond of the dative structure by a high dielectric environment. It can be thought that the polymer plays a role in giving a low dielectric environment. A polymer micelle also provides the low dielectric environment. In addition, the polymeric legand gives the following effects: The addition of salt makes the quaternized poly(4-vinylpyridine) (QPVP) – heme complex shrink, and then k_d is decreased. The addition of sodium dodecylsulfonate extending to CMC and to the pyridine–heme complex decreases k_d. The addition of poly (methacrylic acid) to QPVP–heme complex to form a polymer complex decreases k_d. The lifetime of poly(4-vinyl pyridine-co-N-vinylpyrrolidone) – heme, a non-charged polymer system, is longer than that of QPVP–heme. The higher the hydrophobicity in the polymeric legand, the longer the lifetime of the oxygenated complex. These observations suggest the significance of the hydrophobic polymer environment. In these cases, the Soret band of the oxygenated complex shifts bathochromically and approaches 415 nm which is the absorption band of the oxy-hemoglobin.

Concerning the significance of the hydrophobic polymer environment some examples are shown, such as hydrolyses of PNPA and 8-acetoxyquinoline-5-sulfonate by the acylated cyclodextrin–Ni complex (*161*), cyclodextrin-Cu(II) complex (*162*) and bis-salicylimine of gramicidin-Cu(II) complex (*163*).

Conversely, the importance of the electrostatic environment is indicated in the oxidation reaction by the PLL-Cu(II) complex whose catalytic species is thought to be the planar configuration (*135, 136*).

Usually polymer has its own configuration in the complex formation. The fact that the polymer complex gives a unique polymer radical is shown (*164–169*). Nylon oligomer gives a polymer radical in carbon tetrachloride and it polymerizes vinyl compounds. This initiation reaction of the polymerization reaction was considered as an intramolecular cooperative reaction. The following scheme was presented

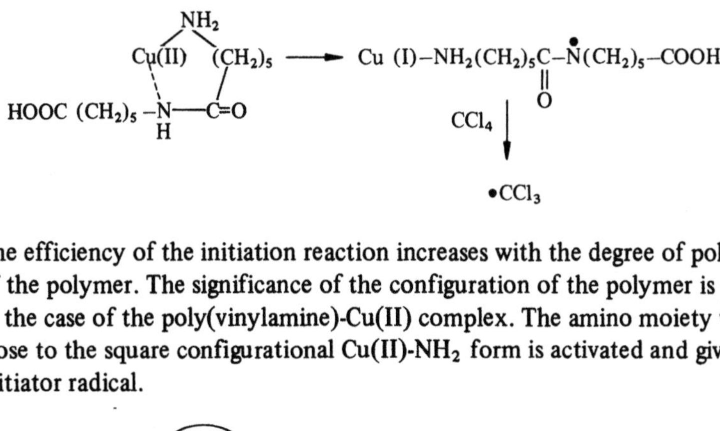

The efficiency of the initiation reaction increases with the degree of polymerization of the polymer. The significance of the configuration of the polymer is pointed out in the case of the poly(vinylamine)-Cu(II) complex. The amino moiety which is close to the square configurational Cu(II)-NH_2 form is activated and gives the initiator radical.

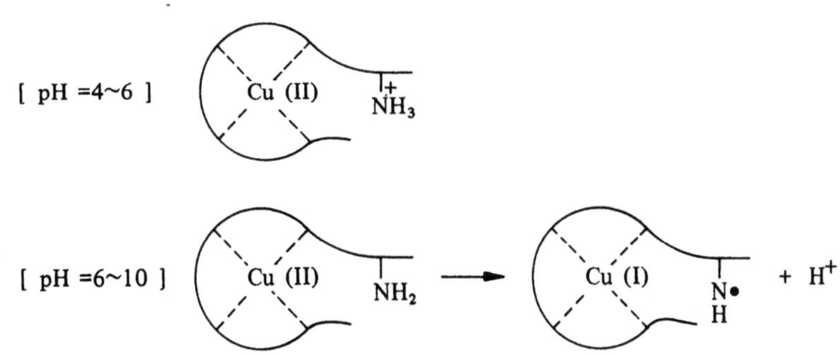

5. Conclusions and Outlook

Many nucleophile-containing polymers have been synthesized as the concrete or the abstract materialization of functional biopolymers. They have their own characteristic features. Some of them follow enzymes and the others share their own chemical fields. To date they have shown their special functions resulting from cooperations. The cooperation in the polymeric system is given by statistical and macroscopic description; however, the detailed study and a microscopic approach gives a possibility of the molecular level discussion. One of the most significant roles of the polymer is to give a specific reaction environment to the reaction field which is reflected in the nature of the polymer. The specific environment given by the polymer makes the nucleophilicity of the nucleophile control and the reaction in a moderate reaction condition possible. Also, it is necessary to analyse the microstructure of the effective catalytic site to materialize the desired functional polymer. In any case, we have by now several functional polymers whose activities are higher than the enzyme in quantity, without selectivity. The selectivity is also one of the more important problems.

6. References

1. Blow, D. M., Birktoft, J. J., Hartley, B. S.: Nature **221**, 337 (1969).
2. Henderson, R.: J. Mol. Biol. **54**, 341 (1970).
3. Shiao, D. F.: Biochemistry **9**, 1083 (1970).
4. Lumry, R., Rajander, S.: Biopolymers **9**, 1125 (1970).
5. Fersht, A. R., Sperling, J.: J. Mol. Biol; **74**, 137 (1973).
6. Robillard, G., Shulman, R. C.: J. Mol. Biol. **71**, 507 (1972).
7. Umeyama, H., Imamura, A., Nagata, C., Hanano, M.: J. Theoret. Biol. **41**, (1973).
8. Gutfreund, H., Sturtevant, J. M.: Biochem. J. **63**, 656 (1956).
9. Bender, M. L.: J. Amer. Chem. Soc. **84**, 2582 (1962).
10. Bender, M. L., Ketzdy, F. J., Gunter, C. R.: J. Amer. Chem. Soc. **86**, 3714 (1964).
11. Lipscomb, W. N., Reeke, Jr., G. N., Hartsuck, J. A., Quiccho, F. A.: Proc. Nat. Acad. Sci. **64**, 28 (1969).
12. Griffith, J. S.: Proc. Royal. Soc. **A235**, 23 (1956).
13. Pauling, L.: Nature **203**, 1182 (1964).
14. Weiss, J.: Nature **202**, 83 (1964).
15. Wüthrich, K.: Structure and Bonding **8**, 53 (1970).
16. Iizuka, T., Morishima, I.: Biochim. Biophys. Acta. **371**, 1 (1974).
17. Morishima, I., Iizuka, T.: J. Amer. Chem. Soc. **96**, 5297 (1974).
18. Sheehan, J. C., Bennett, G. B., Schneider, J. A.: J. Amer. Chem. Soc. **88**, 3455 (1966).
19. Katchalski, E., Fasman, G. D., Simons, E., Blut, E. R., Gurd, F. R. N., Koltun, W. L.: Arch. Biochem. biophys. **88**, 361 (1960).
20. Noguchi, J., Saito, T.: Polyamino acids, polypeptides and proteins (ed. M. A. Stahmann), p. 313. Madison, Wis.: University of Wisconsin Press 1962.
21. Overberger, C. G., Vorchheimer, N.: J. Amer. Chem. Soc. **85**, 951 (1963).
22. Overberger, C. G., St. Pierre, T., Yaroslavsky, C., Yaroslavsky, S.: J. Amer. Chem. Soc. **86**, 1184 (1966).
23. Overberger, C. G., St. Pierre, T., Vorchheimer, N., Lee, J., Yaroslavsky, S.: J. Amer. Chem. Soc. **87**, 296 (1965).
24. Overberger, C. G., Morimoto, M.: J. Amer. Chem. Soc. **93**, 3222 (1971).
25. Overberger, C. G., Morimoto, M., Cho, I., Salamone, J. C.: J. Amer. Soc. **93**, 3228 (1971).
26. Overberger, C. G., St. Pierre, T., Yaroslavsky, S.: J. Amer. Chem. Soc. **87**, 4310 (1965).
27. Kopple, K. D.: Biopolymers **6**, 1417 (1968).
28. Overberger, C. G., Salamone, J. C., Yaroslavsky, S.: J. Amer. Chem. Soc. **89**, 6231 (1967).
29. Shimidzu, T., Furuta, A., Watanabe, T., Kato, S.: Makromol. Chem. **175**, 119 (1974).
30. Overberger, C. G., Maki, H.: Macromolecules **3**, 214 (1970).
31. Overberger, C. G., Maki, H.: Macromolecules **3**, 220 (1970).
32. Shimidzu, T., Furuta, A., Nakamoto, Y.: Macromolecules **7**, 160 (1974).
33. Kunitake, T., Shinkai, S.: Makromol. Chem. **151**, 127 (1972).
34. Kunitake, T., Shinkai, S.: Bull. Chem. Soc. Japn. **43**, 1109 (1970).
35. Kunitake, T., Shinkai, S.: J. Amer. Chem. Soc. **93**, 4256 (1971).
36. Kunitake, T., Shimada, F., Aso, C.: J. Amer. Chem. Soc. **91**, 2716 (1969).
37. Kunitake, T., Shinkai, S.: J. Amer. Chem. Soc. **93**, 4247 (1971).
38. Kunitake, T., Okahata, Y.: Chem. Lett **1974**, 1057.
39. Kunitake, T., Okahata, Y., Ando, R.: Bull. Chem. Soc. Japan **47**, 1509 (1974).
40. Kunitake, T., Okahata, Y., Ando, R.: Macromolecules **7**, 140 (1974).
41. Letsinger, R. L., Savereide, T. J.: J. Amer. Chem. Soc. **84**, 3122 (1962).
42. Overberger, C. G., Yuen, P. S.: J. Amer. Chem. Soc. **92**, 1667 (1970).
43. Morawetz, H., Overberger, C. G., Salamone, J. C., Yaroslavsky, S.: J. Amer. Chem. Soc. **90**, 651 (1968).
44. Michaelis, L., Menten, M. L.: Biochem. Z. **49**, 333 (1913).
45. Langmuir, I.: J. Amer. Chem. Soc. **40**, 1316 (1918).
46. Letsinger, R. L., Klaus, I. L.: J. Amer. Chem. Soc. **86**, 3884 (1964).

47. Letsinger, R. L., Klaus, I. L.: J. Amer. Chem. Soc. **87**, 3380 (1965).
48. Pecht, I., Levitzki, A., Anbar, M.: J. Am. Chem. Soc. **89**, 1587 (1969).
49. Kunitake, T., Shimada, F., Aso, C.: Makromol. Chem. **126**, 276 (1969).
50. Molyneux, P., Frank, H. P.: J. Amer. Chem. Soc. **83**, 3169 (1961).
51. Katchalsky, A., Spitnik, P.: J. Polym. Sci. **2**, 432 (1947).
52. Kauzmann, W.: Adv. Protein. Chem. **14**, 1 (1959).
53. Pshezhetskii, V. S., Murutazaeva, G. A., Kabanov, V. A.: Europ. Polymer J. **10**, 739 (1974).
54. Overberger, C. G., Smith, T. W.: Macromolecules **8**, 401 (1975).
55. Overberger, C. G., Smith, T. W.: Macromolecules **8**, 407 (1975).
56. Overberger, C. G., Morimoto, M., Cho, I., Salamone, J. C.: Macromolecules **2**, 553 (1969).
57. Overberger, C. G., Glowaky, R. C., Vandewyer, P.-H.: J. Amer. Chem. Soc. **95**, 6008 (1973).
58. Sakurada, I., Sakaguchi, Y., Ono, T., Ueda, T.: Makromol. Chem. **91**, 243 (1966).
59. Yoshikawa, S., Kim, O.-K.: Bull. Chem. Soc. Japn. **39**, 1515 (1966).
60. Sakurada, I.: J. Pure Appl. Chem. **16**, 236 (1968).
61. Klotz, I. M., Stryker, V. H.: J. Amer. Chem. Soc. **90**, 2717 (1968).
62. Royer, G. P., Klotz, I. M.: J. Amer. Chem. Soc. **91**, 5885 (1969).
63. Kirsh, Yu. E., Kabanov, V. A., Kargin, V. A.: Vysokomol. Soed. **A10**, 349 (1968).
64. Kirsh, Yu. E., Pluzhnov, S. K., Shomina, T. S., Kabanov, V. A., Kargin, V. A.: Vysokomol. Soed. **A12**, 186 (1970).
65. Imanishi, Y., Sugihara, T., Tanihara, M., Higashimura, T.: Chem. Lett. **1975**, 261.
66. Kunitake, T., Shinkai, S.: Polymer J. **4**, 253 (1973).
67. Overberger, C. G., Podsiadly, C. J.: Bioorg. Chem. **3**, 16 (1974).
68. Overberger, C. G., Podsiadly, C. J.: Bioorg. Chem. **3**, 35 (1974).
69. Okubo, T., Ise, N.: J. Org. Chem. **38**, 3120 (1973).
70. Overberger, C. G., Salamone, J. C.: Accounts Chem. Res. **2**, 217 (1969), and the related references cited therein.
71. Bell, R. P., Jensen, M. J.: Proc. Royal Soc. **A261**, 38 (1961).
72. Overberger, C. G., Pacansky, T. J.: J. Polym. Sci. Polym. Chem. Ed. **13**, 931 (1975).
73. Rony, P. R.: J. Amer. Chem. Soc. **90**, 2824 (1968).
74. Bruice, T. C., Herz, J. L.: J. Amer. Chem. Soc. **86**, 4109 (1964).
75. Bruice, T. C., Schmir, G. L.: J. Amer. Chem. Soc. **79**, 1663 (1957).
76. Bender, M. L., Turnquest, B. W.: J. Amer. Chem. Soc. **79**, 1656 (1957).
77. Overberger, C. G., Salamone, J. C. Yaroslavsky, S.: J. Amer. Chem. Soc. **89**, 6231 (1967).
78. Overberger, C. G., Salamone, J. C., Yaroslavsky, S.: Pure Appl. Chem. **15**, 453 (1967).
79. Shimidzu, T., Chiba, H., Yamazaki, K., Minato, T.: Macromolecules, No. 4 (1976), in press.
80. Katchalski, E., Fasman, G. D., Simons, E., Blout, E. R., Gurd, F. R. N., Koltun, W. L.: Arch. Biochem. Biophys. **88**, 361 (1960).
81. Cruickshank, P., Sheehan, J. C.: J. Amer. Chem. Soc. **86**, 2070 (1964).
82. Kunitake, T., Okahata, Y.: Bioorg. Chem. **4**, 136 (1975).
83. Overberger, C. G., Glowacky, R. C.: J. Amer. Chem. Soc. **95**, 6014 (1973).
84. Johnson, T. W., Klotz, I. M.: Macromolecules **6**, 788 (1973).
85. Böhmer, V., Sauerberry, H.-E., Kämmerer, H.: Makromol. Chem. **175**, 391 (1974).
86. Böhmer, V., Sauerberry, H.-E., Kämmerer, H.: Makromol. Chem. **175**, 2505 (1974).
87. Ergel, A., Bünemann, H., Dattagupta, N.: Makromol. Chem. **175**, 1847 (1974).
88. Kunitake, T., Okahata, Y., Sakamoto, T.: Chem. Lett. **1975**, 459.
89. Overberger, C. G., Okamoto, Y.: Macromolecules **5**, 363 (1972).
90. Kabanov, V. A., Kirsh, Yu. E., Papisov, I. M., Torcilin, V. P.: Vysokomol. Soed. **B14**, 405 (1972).
91. Bruice, T. C., Schmir, G. L.: J. Amer. Chem. Soc. **80**, 148 (1958).
92. Blow, D. M., Steiz, T. A.: Ann. Rev. Biochem. **1970**, 716.
93. Kirsch, J. F., Jencks, W. P.: J. Amer. Chem. Soc. **85**, 837 (1964).
94. Kunitake, T., Shinkai, S., Hiroyasu, S.: J. Polym. Sci., Polym. Lett. Ed., in press.
95. Kirsh, Yu. E., Rahnanskaya, A. A., Lukovkin, G. M., Kabanov, V. A.: Europ. Polym. J. **10**, 393 (1974).
96. Yatsimirski, A. K., Martinek, K.: Berezin, IV. Tetrahedron **27**, 2855 (1971).

97. Tagaki, W., Chihiro, M., Amada, T., Yano, Y.: Chem. Comm. **1972**, 219.
98. Tagaki, W., Amada, T., Yamashita, Y., Yano, Y.: Chem. Comm. **1972**, 1131.
99. Tabushi, I., Kuroda, Y., Kita, S.: Tetrahedron Lett. 643 (**1974**).
100. Tabushi, I., Kuroda, Y.; Tetrahedron Lett. 3613 (**1974**).
101. Kunitake, T., Okahata, Y., Sakamoto, T.: Chem. Lett. **1975**, 459.
102. Suzuoki, K., Imanishi, Y., Higashimura, T., Okamura, S.: Biopolymers 7, 917 (1969).
103. Suzuoki, K., Imanishi, Y., Higashimura, T., Okamura, S.: Biopolymers 7, 925 (1969).
104. Nozawa, T., Hatano, M., Kambara, S.: Makromol. Chem. **112**, 73 (1968).
105. Hatano, M., Nozawa, T., Yamamoto, T., Kambara, S.: Makromol. Chem. **115**, 10 (1968).
106. Pecht, I., Levitzki, A., Amber, M.: J. Amer. Chem. Soc. **89**, 1587 (1967).
107. Levitzki, A., Pecht, I., Berger, A.: J. Amer. Chem. Soc. **94**, 6844 (1972).
108. Nozawa, T., Hatano, M., Kambara, S.: Kogyokagaku Zasshi 72, 369 (**1969**).
109. Siegel, H., Brauer, G.: Helv. Chim. Acta. **51**, 1246 (1969).
110. Tsuchida, E., Honda, K., Sato, H.: Biopolymers **13**, 2147 (1974).
111. Shirai, H., Inaki, Y., Takemoto, K.: Makromol. Chem. **175**, 2047 (1974).
112. Grubbs, R. H., Gibbons, C. G., Kroll, L. C., Bonds, W. D., Brubaker, C. H.: J. Amer. Chem. Soc. **95**, 2373 (1973).
113. Capka, M., Svoboda, P., Cerny, M., Hetflejs, J.: Tetrahedron Lett. 4787 (**1971**).
114. Grubbs, R. H., Kroll, L. C., Sweat, E. W.: J. Macromol. Sci. Chem. **A7**, 1047 (1973).
115. Iwasawa, Y., Fujitsu, H., Onishi, T., Tamaru, K.: J. Chem. Soc. FaradayI, **70**, 202 (1974).
116. Nakamura, Y., Hirai, H.: Chem. Lett. **1974**, 645.
117. Nakamura, Y., Hirai, H.: Chem. Lett. **1974**, 809.
118. Bruner, H., Bailer, J. C.: Inorg. Chem. **12**, 1645 (1973).
119. Tsuchida, E., Nishide, H., Nishiyama, T.: J. Polym. Sci. **C45**, 248 (1974).
120. Tsuchida, E., Nishide, H., Nishiyama, T.: Makromol. Chem. **176**, 1349 (1975).
121. Tsuchida, E., Kaneko, M., Nishide, H.: Makromol. Chem. **164**, 203 (1973).
122. Tsuchida, E., Nishide, H., Takeshita, M.: Makromol. Chem. **175**, 2293 (1974).
123. Tsuchida, E., Karino, H., Nishide, H., Kurimura, Y.: Makromol. Chem. 161 (**1974**).
124. Tsuchida, E., Karino, H., Nishide, H., Kurimura, Y.: Makromol. Chem. **175**, 171 (1974).
125. Gould, E. S.: J. Amer. Chem. Soc. **92**, 6797 (1970).
126. Vogel, B., Morawetz, H.: J. Amer. Chem. Soc. **90**, 1368 (1968).
127. Morawetz, H., Vogel, B.: J. Amer. Chem. Soc. **91**, 563 (1969).
128. Morawetz, H., Gordimer, G.: J. Amer. Chem. Soc. **92**, 7532 (1970).
129. Ise, N.: Nature **225**, 66 (1970).
130. Ise, N., Matsuda, Y.: J. Chem. Soc. Faraday I **69**, 99 (1973).
131. Breslow, R., Overman, L. E.: J. Amer. Chem. Soc. **92**, 1075 (1970).
132. Lautsh, W., Broser, W., Biedelman, W., Grichtel, H.: J. Polymer Sci. **17**, 479 (1955).
133. Wang, J. H., Bringer, W. S.: Proc. Natl. Acad. Sci. U. S. **46**, 958 (1962).
134. Moriguchi, Y.: Bull. Chem. Soc. Japn. **39**, 2656 (1966).
135. Hatano, M., Nozawa, T., Ikeda, S., Yamamoto, T.: Makromol. Chem. **141**, 11 (1971).
136. Nozawa, T., Hatano, M.: Makromol. Chem. **141**, 31 (1971).
137. Nozawa, T., Akimoto, Y., Hatano, M.: Makromol. Chem. **158**, 21 (1972).
138. Lautsch, W., Schroder, E.: Monath. Chem. **88**, 432 (1957).
139. Welch, R. C. W., Rase, H. F.: Ind. Eng. Chem. Fundamentals **8**, 389 (1969).
140. Hirai, H., Furuta, T.: J. Polym. Sci. **B9**, 459 (1971).
141. Hirai, H., Furuta, T.: J. Polym. Sci. **B9**, 729 (1971).
142. Beamer, R. L., Fickling, C. S., Eming, J. H.: J. Pharm. Sci. **56**, 1029 (1967).
143. Beamer, R. L., Belding, R. H., Fickling, C. S.: J. Pharm. Sci. **58**, 1142 (1969).
144. Beamer, R. L., Belding, R. H., Fickling, C. S.: J. Pharm. Sci. **58**, 1142 (1969).
145. Plate, N. A., Davidova, S. L., Alieva, E. D., Kargin, V. A.: Europ. Polymer J. **6**, 1371 (1970)
146. Harada, K., Yoshida, T.: Naturwiss. **57**, 131 (1970).
147. Harada, K., Yoshida, T.: Naturwiss. **57**, 306 (1970).
148. Collman, J. P., Reed, C. A.: J. Amer. Chem. Soc. **95**, 204 (1973).
149. Collman, J. P., Gagne, R. R., Kouba, J., Wahren, H. L.: J. Amer. Chem. Soc. **96**, 6800 (1974).

150. Kamogawa, H.: J. Polym. Sci. **12**, 2317 (1974).
151. Rollman, L. D.: J. Amer. Chem. Soc. **97**, 2132 (1975).
152. Leal, O., Anderson, D. L., Bowman, R. G., Basolo, F., Burwell, Jr, R. L.: J. Amer. Chem. Soc. **97**, 5125 (1975).
153. Grubbs, R. H., Kroll, L.: J. Amer. Chem. Soc. **93**, 3062 (1971).
154. Pittman, Jr, C. U., Smith, L. R.: J. Amer. Chem. Soc. **97**, 341 (1975).
155. Pittman, Jr, C. U., Hanes, R. M.: J. Macromol. Sci. **7**, 1563 (1973).
156. Pittman, Jr, C. U., Hanes, R. M.: Macromolecules **7**, 396 (1974).
157. Kim, S. J., Takizawa, T.: Makromol. Chem. **175**, 125 (1974).
158. Tsuchida, E., Honda, K., Sata, H.: Biopolymers **13**, 2147 (1974).
159. Tsuchida, E., Honda, K.; Polymer J. **7**, 498 (1975).
160. Tsuchida, E., Honda, K., Hasegawa, E.: Biochem. Biophys. Acta **393**, 483 (1975).
161. Breslow, R., Overman, L. E.: J. Amer. Chem. Soc. **92**, 1095 (1970).
162. Matsui, Y., Toyoaki, K., Date, Y.: Bull. Chem. Soc. Japn. **45**, 3229 (1972).
163. DeSantis, P., Delario, L., Lamana, G.: Biopolymer **12**, 423 (1973).
164. Takada, T., Takemoto, K.: Angew. Makromol. Chem. **19**, 1 (1971).
165. Takemoto, K., Takada, T., Inaki, Y.: J. Polymer Sci. **A1**, 1061 (1972).
166. Takemoto, K., Inaki, Y.: Prog. Polymer Sci. Japn. **8**, 43 (1975).
167. Inaki, Y., Kimura, K., Takemoto, K.: Makromol. Chem. **171**, 19 (1973).
168. Kimura, K., Inaki, Y., Takemoto, K.: Makromol. Chem. **175**, 83 (1974).
169. Kimura, K., Inaki, Y., Takemoto, K.: Makromol. Chem. **175**, 95 (1974).
170. Baumgartner, E., Fernández-Prini, R., Turyn, D.: J. Chem. Soc. Faraday **I70**, 1518 (1974).

Received March 8, 1976

The Formation of Cyclic Oligomers in the Cationic Polymerization of Heterocycles

Eric J. Goethals
Institute of Organic Chemistry, Rijksuniversiteit Gent, Krijgslaan 271, 9000 Gent, Belgium

Table of Contents

A) Introduction

According to a number of reports of the last two decades it now has become more and more evident that cationic polymerization of heterocyclic monomers may lead to polymers containing more or less considerable amounts of cyclic oligomers. In many cases even, the oligomers are the main reaction products. If the polymers are the desired end-products, oligomer formation must be regarded as an undesired side reaction. Therefore, it is useful to know by which mechanism these oligomers are formed in order to be able to conduct the experiments under such conditions that this side reaction is minimized. On the other hand, it may be equally interesting to influence the reaction in such a manner that the cyclic oligomers become the main reaction products because this can lead to new and relatively easy methods for the synthesis of heterocyclic compounds which would otherwise be difficult to prepare. This is especially the case when macrocyclic oligomers are formed since during last years the usefulness of such compounds (crown ethers) has well been recognized (*1*).

The aim of this article is to review the different systems where such reactions have been observed, to discuss some of the proposed mechanisms for oligomer formation and finally to try to derive some general rules which govern this kind of reactions.

B) Survey of Oligomer-Forming Monomers

1. Oxiranes

The best known example of an oligomer-forming monomer is ethylene oxide. This monomer has been studied extensively by Eastham and coworkers (*2–4*). When ethylene oxide is polymerized by borontrifluoride in 1,2-dichloroethane, the polymerization proceeds only to a molecular weight of about 700, although monomer continues to disappear more or less indefinitely, but this is due almost entirely to the formation not of new polymer but of dioxane (*3*). It is believed that the dioxane is formed by degradation of the polymer via oxonium salts (*4*). This was deduced from the observation that, if high molecular weight polyglycols are introduced to the reaction mixture, their molecular weight is reduced until the average molecular weight is once again about 700. This reduction in molecular weight is accomplished by elimination of dioxane from the polymer chain as is shown by the fact that on addition of deuterated ethylene oxide to a polymer, undeuterated dioxane is the only reaction product at the early stages of the depolymerization. Triethyloxonium tetrafluoroborate degrades polyethyleneglycol to dioxane and attempts to polymerize ethylene oxide with this oxonium salt failed, the only reaction product being dioxane. Apparently, dioxane formation via oxonium salts is more favored than propagation.

Figure 1 shows the results obtained when samples of triethyloxonium tetrafluoroborate are allowed to react to completion with a series of polyglycols in 1,2-dichloroethane solution. Equal weights (0.86 g) of polymer were taken for all experiments so the essential difference between runs with polymers of molecular weight 1540, 4000, 6000 and 20.000 lies in the hydroxyl content of the solutions. With sufficient oxonium salt the polymers were completely degraded and the only reaction product was dioxane.

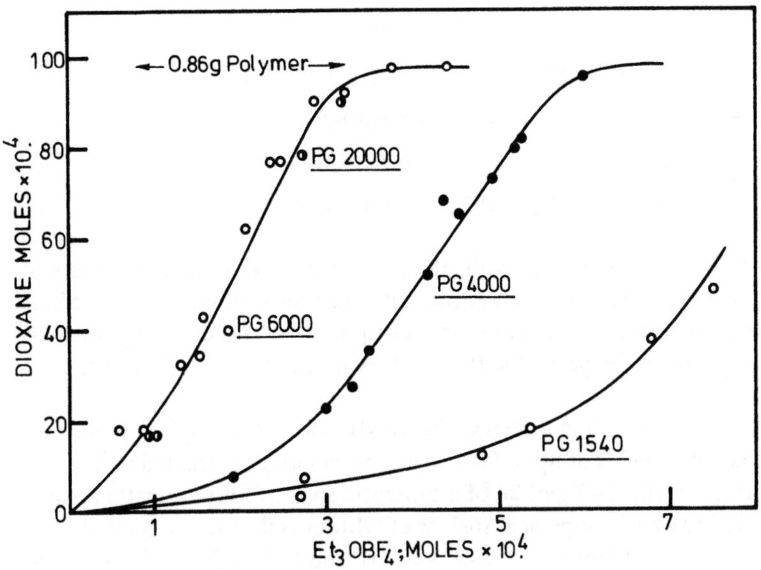

Fig. 1. Depolymerization of polyglycols by triethyloxonium tetrafluoroborate (*4*). Solv.: ethylene dichloride, temp. 20 °C. In the curve on the extreme left, o corresponds to MW 6.000 and ◑ to MW 20.000

In these experiments with oxonium salts no 2-methyl-1,3-dioxolane was formed as was the case in the experiments with ethylene oxide and borontrifluoride. The effectiveness of the oxonium salt in decomposing the polymer increases with increasing molecular weight of the polymer (decreasing hydroxyl content) until an apparent maximum is reached at a molecular weight of 6000. This suggests that the formation of dioxane from the polymer is a chain reaction that is stopped by two different termination reactions: a first one occurs with hydroxyl groups and another occurs when an average of forty molecules of dioxane have been formed per molecule of initiator. The mechanism of the degradation proposed by Eastham is as follows: The first step is an ether exchange reaction which transfers the oxonium ion to the polyglycol chain:

$$R_3O^{\oplus} + \sim OCH_2CH_2OCH_2CH_2O \sim \;\rightarrow\; R_2O + \sim OCH_2CH_2\underset{\displaystyle R}{\overset{\oplus}{O}}-CH_2CH_2O \sim$$

Internal ether exchange would then eliminate dioxane until the oxonium ion is destroyed in one or both of the following termination steps.

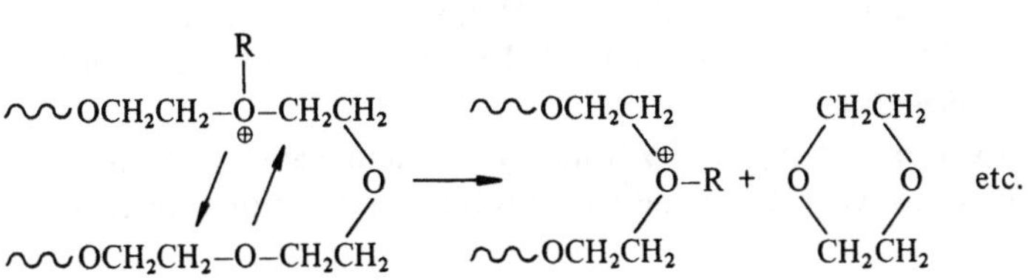

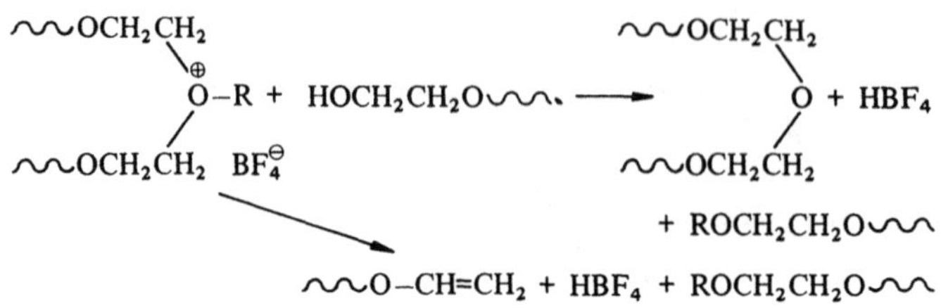

Polymerization of ethylene oxide with stannic chloride also leads to a mixture of polymer and dioxane (*2*). However, in contrast with the borontrifluoride initiated reaction the polymers can reach a molecular weight of up to 20.000 and chain growth and dioxane formation seem to be parallel with each consuming about one half of the monomer.

In a recent paper Saegusa and his co-workers (*64*) report that ethylene oxide is converted to dioxane in yields of up to 96% when the monomer is treated with a catalytic amount (generally 2–5 mol %) of a superacid such as trifluoromethanesulfonic acid, or a derivative of a superacid such as ethylfluorosulfonate, in methylene chloride or nitromethane at temperatures between 10 and 40 °C. The authors propose that dioxane is formed by a simultaneous polymerization and degradation of the formed polymer. Propagation as well as degradation are assumed to occur via the ester species.

For example, with ethyltrifluoromethyl sulfonate as the initiator:

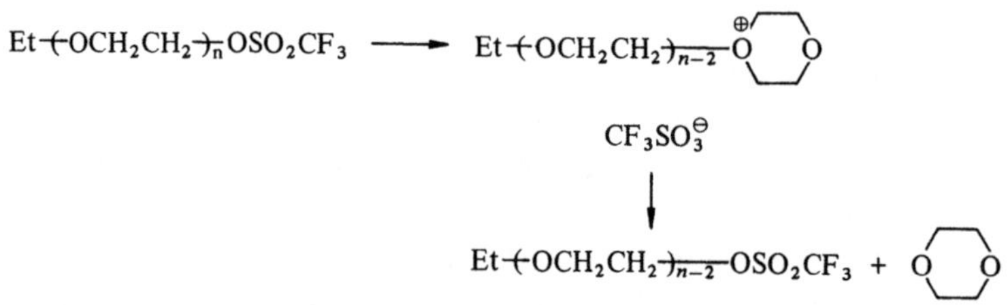

Recently it has been reported (*5*, *66*) that ethylene oxide leads to a mixture of cyclic oligomers of D. P. 3, 4, 5, 6, 7, 8 and 9 (7:4:6:6:2:1), unaccompanied by open-chain oligomers and polymers, when treated with fluorine-containing catalysts such as antimony pentafluoride or a 1:1 mixture of borontrifluoride and hydrogen fluoride.

A number of other epoxides have been found to produce cyclic oligomers. As shown in Table 1, propylene oxide, when treated with triethyloxonium tetrafluoroborate or borontrifluoride, produces a mixture of cyclic oligomers, the most important one being the tetramer (*6*, *7*).

Because of the occurrence of head-to-tail and head-to-head structures and of *cis-trans* isomerism the cyclic tetramer of propylene oxide can occur as 23 different geometrical isomers. By a combination of gas chromatography, H^1-NMR- and mass spectroscopy, Katnik and Schaefer (*6*) were able to observe 22 out of the 23 isomers and to characterize the majority of them.

When propylene oxide is polymerized by trimethyl aluminum at 0 °C, cyclic tetramer together with some cyclic dimer and trimer is formed in the first stages of the polymerization (*8*).

Table 1. Reaction products obtained from propylene oxide (166 g) and triethyloxonium tetrafluoroborate (40 mmoles) (*7*)

Boiling range, °C	Amount, g	Principal species[a]
117–130 (1 atm)	7	$(PO)_2$
6– 65 (0,07 mm)	67	$(PO)_4$
77– 83 (0,07 mm)	15	$(PO)_5$
90–150 (0,07 mm)	6	
residue	36	Resin

[a] PO = $CH(CH_3)CH_2O$

Polymerization of 1,2-butylene oxide with triethyloxonium tetrafluoroborate or borontrifluoride leads to a mixture of polymer, dimer, tetramer and larger rings (*7*). The cyclic tetramer is the most important reaction product.

The formation of small amounts of 2,5-diphenyl dioxane from styrene oxide has been reported by Colclough and co-workers (*9*), and by Pasika (*10*). The first authors used Friedel Crafts catalysts and the maximal yield of dimer was 10%. (The same catalysts gave no oligomers with ethylene oxide or propylene oxide). The second author obtained the cyclic dimer in 2–3% yield when styrene oxide was polymerized with triphenylmethyl hexachloroantimonate. Pasika's dimer has a m.p. of 171–172 °C and the author presents good evidence that it is the *trans*-isomer. Colclough and co-workers do not discuss the geometry of their dimer and report a sharp m.p. of 152 °C. Possibly, the lower melting compound is the *cis*-isomer or a mixture of *cis*- and *trans*-isomers. *Trans*-2,5-diphenyl dioxane has also been isolated by Kondo and Blanchard (*11*) during the cationic copolymerization of styrene oxide with propylene oxide. These authors report a m.p. of 177 °C.

In a review article Eastham (*12*) reports that treatment of isobutylene oxide with borontrifluoride leads to dimer without any polymer but the reaction conditions were not mentioned. On the other hand, Vandenberg (*13*) described the polymerization of the same monomer with borontrifluoride to give 60% of polymer containing substantial amounts (ca. 27%) of high boiling volatiles which were supposed to be a mixture of cyclic oligomers.

The formation of cyclic oligomers from epichlorohydrin has been reported in several papers (*7, 14–17*). Weissermel and Nölken (*14*) reported that at low temperature epichlorohydrin (ECH) reacts with triethyl aluminum to form a mixture of polymer and the cyclic dimer. The proposed mechanism is as follows (see page 108).

Kern (*7*) and Entelis and co-workers (*15–17*) found tetramer and small amounts of larger rings when the monomer was treated with borontrifluoride or triethyloxonium tetrafluoroborate. With the first initiator the ratio of tetramer to polymer is greater in the early stages of the reaction than after longer reaction periods (*15*). From this observation it was concluded that tetramer and polymer are formed by two independent reaction paths. The ratio of oligomers to polymer increased with temperature (*16*).

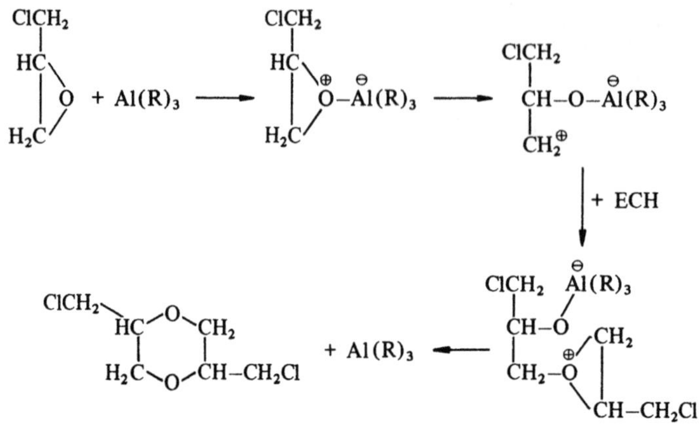

According to Entelis the formation of cyclic oligomers in the polymerization of oxiranes initiated by borontrifluoride would be an intramolecular reaction between the chain ends of the zwitter ion produced in such polymerizations:

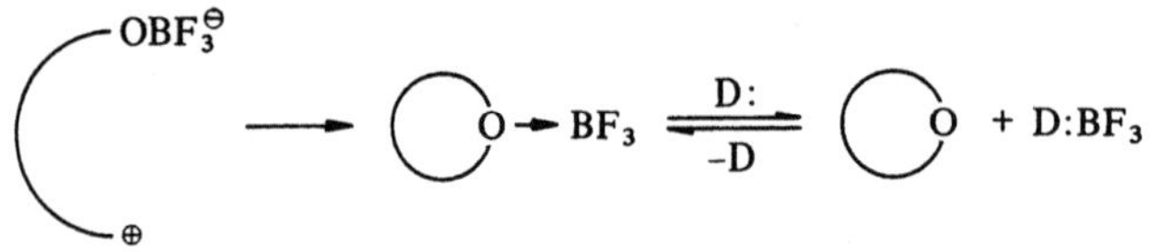

The positive chain end is assumed to be a carbenium ion. This mechanism is not applicable to polymerizations initiated with strong alkylating agents such as triethyloxonium tetrafluoroborate because with these initiators no zwitter ions can be produced. Since no difference was observed between the products resulting from BF_3 or from oxonium initiators (*6*), it is unlikely that tetramer would be formed by two different mechanisms in these two cases and consequently the proposed mechanism is dubious.

The nitrate of glycidyl epoxide forms a mixture of polymer and of cyclic tetramer. The amount of tetramer is smaller than in the cases of epichlorohydrin or propylene oxide which was attributed to the higher steric hindrance in the nitrate (*15*).

Optically active tert.butyl oxirane forms a cyclic tetramer (*18*) with m.p. 168 °C, when treated with borontrifluoride etherate at 20 °C. The yield of tetramer is between 20 and 30%.

2. Oxetanes

In 1956 Rose (*19*) reported that oxetane and 3,3-dimethyl oxetane, when treated with typical cationic initiators, form mixtures of polymer and cyclic tetramer (16-membered rings). Increase of the temperature increases the proportion of tetramer in the reaction products. Oxetane for example gives 4% of tetramer at –80 °C but at 50 °C 66% of the reaction products was tetramer (total tetramer yield based on monomer: 42%). It is not an equilibrium effect, for the final conversion of monomer to polymer and tetramer obtained by heating to 100 °C the product of a polymerization carried out at –80 °C without removing the catalyst, was almost the same as that obtained at –80 °C. The mechanism of tetramer formation proposed by Rose (*19*) is, that after 4 propagation steps the growing species is attacked by the hydroxyl function situated at the beginning of the chain, forming cyclic tetramer and initiator:

$$HO(CH_2)_3\,[-O\,(CH_2)_3]_2 - \overset{\oplus}{O}\,(CH_2)_3 \quad \longrightarrow \quad [O\,(CH_2)_3]_4 + H_2OBF_3$$
$$BF_3OH^{\ominus}$$

However, when triethyloxonium salts are used as initiator, cyclic oligomer is also formed. Since in this case no terminal hydroxyl groups are formed, the formation of the tetramer must also be possible by another reaction mechanism.

From the polymerization products of 3,3-bis(chloromethyl) oxetane with triethyl aluminum a small amount of cyclic trimer has been isolated (*20*).

The oligomer formation during the polymerization of oxetanes has been reininvestigated by two groups recently. Dreyfuss and Dreyfuss (*21*) found that with oxonium salts or with ethyl trifluoromethanesulfonate as initiator, oxetane forms not only cyclic tetramer and polymer as reported by Rose, but also cyclic trimer. The amount and type of oligomer obtained depends on counter ion, temperature and solvent. Polymeri-

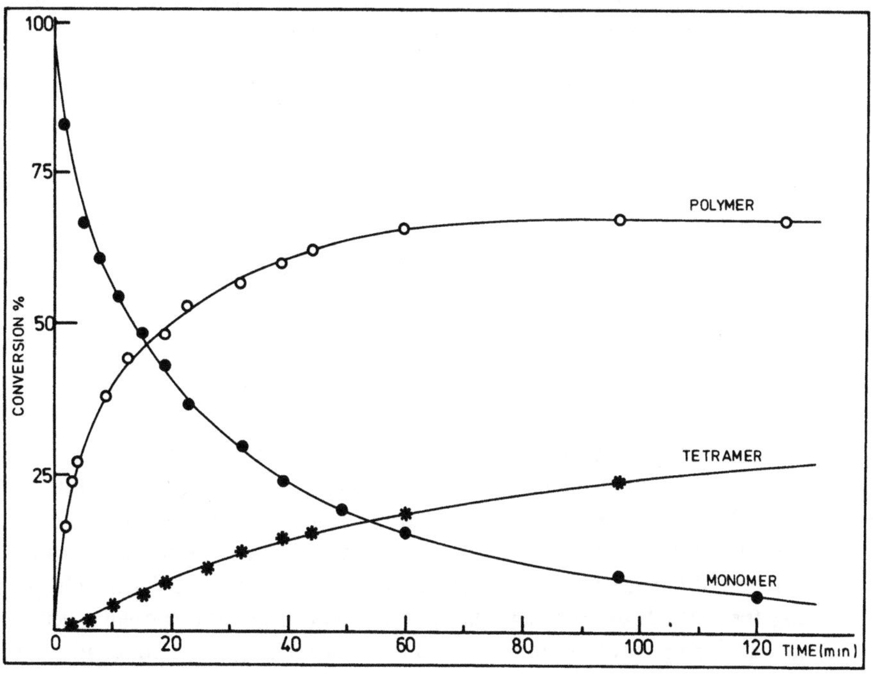

Fig. 2. Polymerization of 3,3-dimethyl oxetane with triethyloxonium tetrafluoroborate: conversion of monomer into polymer and cyclic tetramer in function of time (*22*). Solv.: methylene chloride, temp. 18 °C
initial monomer conc.: 0.73 mol · l^{-1}, initiator conc.: 0.03 mol · l^{-1}

zation initiated by triethyloxonium tetrafluoroborate resulted in as much as 35% conversion to oligomer at complete monomer conversion. Cyclic tetramer was predominant. Other oxonium salts (PF_6^-, SbF_6^- and $SbCl_6^-$) produced insignificant amounts of cyclic oligomers. Polymerizations initiated by ethyl trifluoromethanesulfonate led to 15% cyclic oligomer, almost exclusively cyclic trimer. Polymer was not converted to cyclic oligomer by adding initiator to reaction mixtures after complete disappearance of monomer. Cyclic oligomers formed only when the initiator was added to a mixture containing monomer. Therefrom it was concluded that cyclic oligomer formation appears to be a concurrent process with polymerization.

Bucquoye (*22*) made a kinetic study of the formation of cyclic tetramer during the polymerization of 3,3-dimethyloxetane initiated by triethyloxonium tetrafluoroborate. With this monomer also, the oligomer formed only during and not after the polymerization (Fig. 2).

3. Five-, Six- and Seven Membered Cyclic Ethers

The formation of cyclic oligomers in the polymerization of tetrahydrofuran has not been observed (*23*). In the case of 1,3-dioxolane small amounts of oligomers going from dimer to nonamer have been isolated (*24*). Under the influence of borontrifluoride etherate in methylene chloride at 35 °C small amounts of 1,3,5-trioxepane are formed (*25*). Miki, Higashimura and Okamura (*25*) ascribed the formation of this substance to a backbiting reaction in the polymerization of dioxolane:

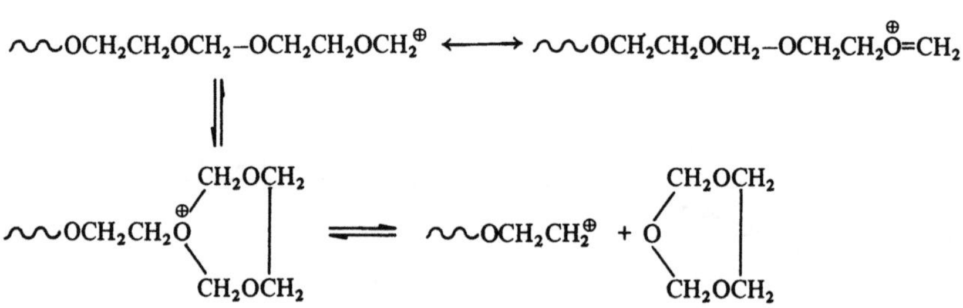

However, because the newly formed carbenium ion is not more stabilized, it is not likely to be formed.

The mechanism of 1,3,5-trioxacycloheptane formation during the polymerization of dioxolane has been reinvestigated by Boehlke and Jaacks (*55*) using borontrifluoride etherate or perchloric acid as initiators. These authors found that the formation of trioxacycloheptane is due to the presence of traces of water. Water causes chain transfer with formation of hemiacetalic hydroxyl end groups which cleave off formaldehyde. The latter together with dioxolane yields the trioxacycloheptane. It is formed quickly in the beginning of the polymerization and then remains unchanged. Its concentration is proportional to the amount of water present in the polymerization mixture. 1,3,5-Trioxacycloheptane is also formed during the copolymerization of trioxane with ethylene oxide at 70 °C, initiated with borontrifluoride (*57*) and during the copolymerization of trioxane and 1,3-dioxolane (*58*).

In a recent article (*59*) Kops and Spanggaard report that *cis*-7,9-dioxabicyclo[4,3,0]-nonane is transformed quantitatively into the cyclic dimer when treated with a catalytic amount of phosphorous pentafluoride or triethyloxonium hexachloroantimonate. From an NMR study the dimer was concluded to have the *cis*-anti-*cis* structure.

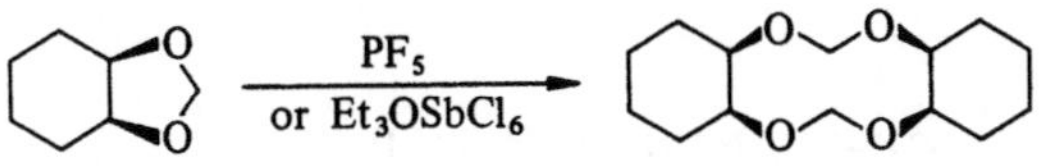

cis–isomer *cis*–anti–*cis* isomer

Trans-7,9-dioxabicyclo[4,3,0]nonane behaved differently, depending on the solvent. In bulk, low molecular weight polymer is formed whereas in 1,2-dichloroethane the reaction product is predominantly the *trans*-anti-*trans* dimer

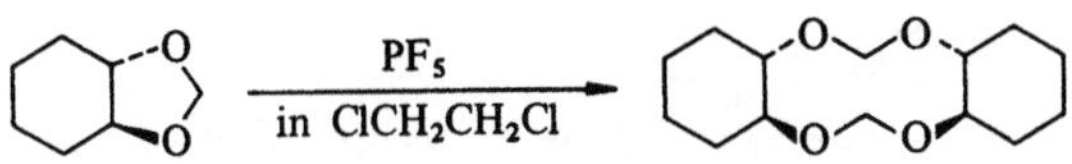

trans–isomer *trans*–anti–*trans*

In toluene with phosphorous pentafluoride as initiator the polymerization was completely reversible with temperature change, leading to a mixture of monomer, cyclic dimer and polymer the ratio of which depended on the temperature. Polymerization starting with the dimer instead of monomer lead to the same equilibrium mixture.

The authors propose that the dimers are formed by a ring-expansion mechanism comparable to the polymerization mechanism for cyclic acetals proposed by Plesch (*30*).

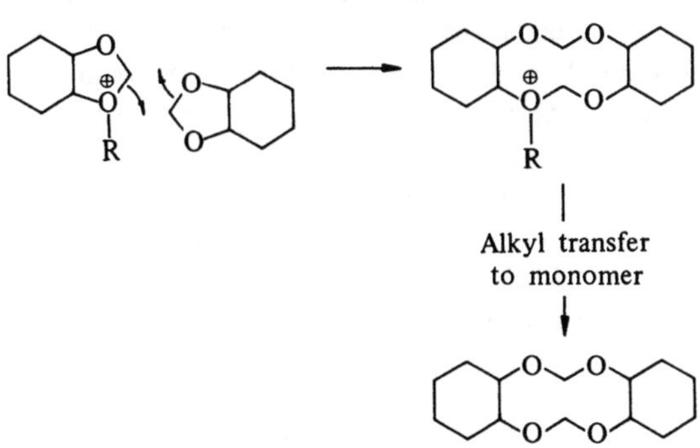

The polymerization of trioxane in solution has been studied by Okamura (*26*) and his co-workers and by Kern and Jaacks (*56, 58, 63*). The initiators were borontrifluoride or its complexes and anhydrous perchloric acid. During the polymerization the eight-membered ring, tetroxane, is formed rapidly but this compound takes part in the polyoxymethylene formation. This results in an equilibrium concentration of tetroxane when the rate of formation becomes equal to the rate of consumption (*27*). Minor amounts of the ten-membered ring, pentoxane, are also formed (*28*). The authors conclude that tetroxane and pentoxane are formed by a back-biting mechanism. It is assumed that the active species in the reaction is a carbenium ion. Although this ion is stabilized by resonance, it is considered to be more unstable than the oxonium ion unless the carbenium ion is stabilized by another molecule. If the carbenium ion attacks a trioxane molecule polymerization proceeds. On the other hand, it may also attack an oxygen atom of its own chain.

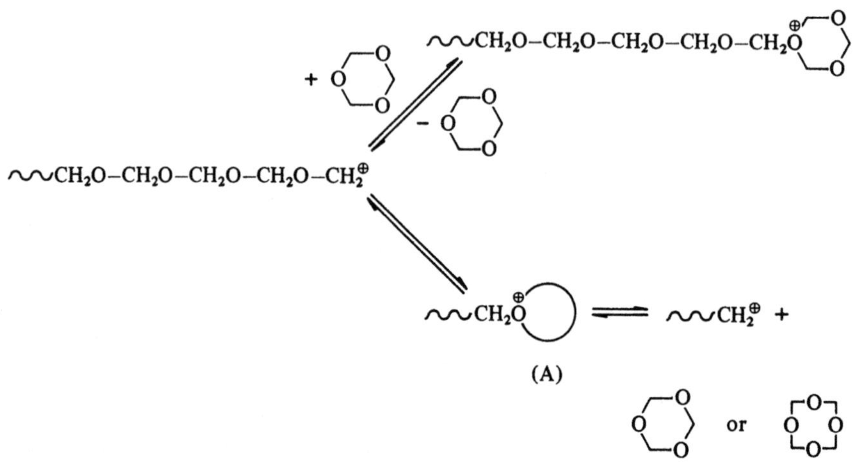

If the carbenium ion attacks the third oxygen atom from the end a six-membered oxonium ion is formed, and if it attacks the fourth, an eight-membered ring is formed. Therefore, the oxonium ion (A) stands for either the six-membered or eight-membered ring. The same reaction mechanism is responsible for the observation that trioxane is produced when a suspension of high molecular weight polyoxymethylene in nitrobenzene is treated with borontrifluoride etherate at 30 °C and also for the formation of 0,22 $mol.l^{-1}$ of trioxane (the ceiling concentration at 25 °C) during the cationic polymerization of formaldehyde in nitrobenzene at 25 °C (*56*).

The occurrence of a back-biting reaction is further indicated by the observation that in copolymerizations of trioxane or tetroxane with vinyl monomers various 1,3-dioxane derivatives are formed (*29, 56*). With styrene for example, the derivative was 4-phenyl-1,3-dioxane, the formation of which was explained as follows:

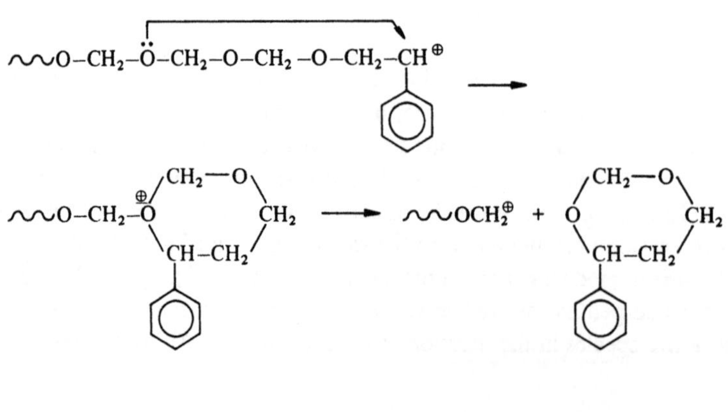

Table 2. Oligomer formation during the cationic polymerization of cyclic ethers

Monomer	Initiator	Solvent	Temp. °C	Structure of oligomer	Ref.
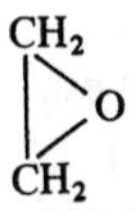	BF_3, $SnCl_4$ Et_3OBF_4	EtCl	20	Dimer	(*2–4*)
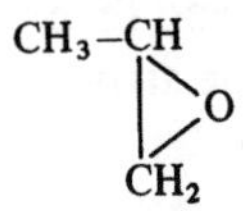	Bf$_3$, Et_3OBF_4	bulk	35	Dimer, tetramer, pentamer	(*6, 7*)
	Al $(CH_3)_3$	CH_2Cl_2	0–80	Dimer, trimer tetramer	(*8*)
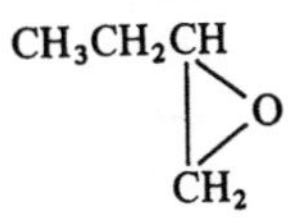	BF_3, Et_3OBF_4	CH_2Cl_2	41	Dimer, tetramer pentamer	(*7*)
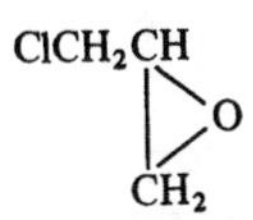	BF_3, Et_3OBF_4	CCl_4	77	Dimer, tetramer, pentamer	(*7*) (*15–17*)
	$AlEt_3$	Et_2O	–50	Dimer	(*14*)
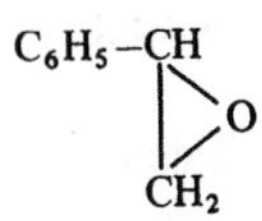	$(C_6H_5)_3C\ SbCl_6$	–	25–50	Dimer	(*10*)
	$AlCl_3$, $FeCl_3$	–	–	Dimer	(*9, 11*)
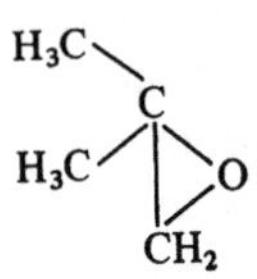	BF_3	–	–	Dimer	(*12*)
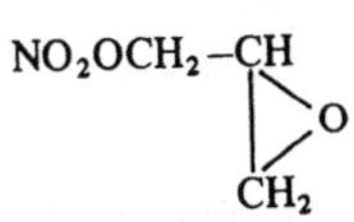	BF_3	–	–	Tetramer	(*15*)
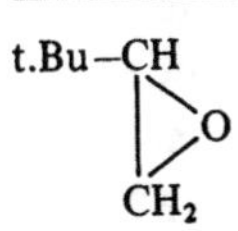	BF_3OEt_2	–	20°	Tetramer	(*18*)
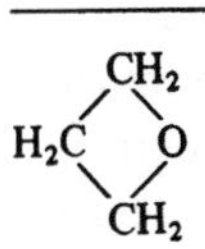	BF_3, Et_3OBF_4	bulk or EtCl	–80–50	Tetramer	(*19, 21*)
	$CF_3SO_3C_2H_5$	C_6H_6	70°	Trimer, tetramer	(*21*)
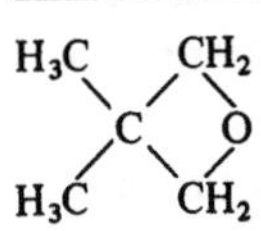	BF_3, Et_3OBF_4	EtCl, CH_2Cl_2	0–40°	Tetramer	(*19, 22*)
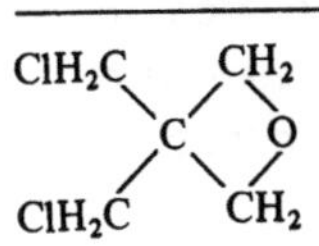	Et_3Al	–	–	Trimer	(*20*)

Plesch and Westermann (*30*) found that under the influence of perchloric acid, 1,3-dioxane does not polymerize but produces a cyclic dimer which was formed only under those conditions under which it could crystallize from the reaction mixture.

The same authors reported that during the polymerization of 1,3-dioxepane with perchloric acid appreciable quantities of dimer are produced (*30*). Yamashita (*31*) who studied the kinetics of the same polymerization, reports that the molecular weight of the polymer increases with conversion until a maximum value is attained and then decreases drastically at higher conversions. Possibly, this phenomenon is to be attributed to a degradation of the polymer to form the dimer or higher oligomers.

Table 2 gives a survey of cyclic oligomer forming oxygen heterocycles.

4. Thiiranes

A series of substituted thiiranes have been investigated in the author's laboratory (*32–35*). All monomers behaved in a similar way: a rapid polymerization was followed by degradation to cyclic oligomers. The rate of the degradation and the ring size of the oligomers varied with the monomer.

Under the influence of triethyloxonium tetrafluoroborate propylene sulfide and 1,2-butylene sulfide polymerize in a 2-stage polymerization which leads to quantitative conversion (*32, 35*). After polymerization the polymers slowly degrade to form a mixture of low molecular weight polymer, a small amount of cyclic pentamer and (approximately 50%) cyclic tetramer (*33*). If high molecular weight poly (propylene sulfide) obtained by an anionic or a coordinative polymerization is treated with a catalytic amount of oxonium salt, the polymer degrades rapidly to form again a mixture of cyclic pentamer and tetramer (*34*) in a ratio of approx. 1 : 10. This confirms that the oligomers are formed by degradation of polymer and provides strong evidence that the active species for the degradation are sulfonium ions (*67*).

In previous papers it was proposed that these oligomers are formed by a simple back biting degradation mechanism (*32–34*). A number of recent observations (*65*) lead to the conclusion that this is not correct: When a new amount of polymer is added to a degrading polymer mixture 15 minutes after the initiation, the added polymer degrades at the same rate as if it were present at the very beginning of the degradation. Since the reaction between the initiator and the polymer is very rapid and quantitative, the active species for the degradation reaction must therefore be transferred very rapidly from one macromolecule to another. When the second amount of polymer was deuterated the cyclic tetramer consisted of a mixture of undeuterated, mono, di, tri and tetra-deuterated products. This is strong evidence that the formation of this oligomer is the result of a very rapid exchange between sulfide functions and sulfonium ions resulting occasionally in the formation of cyclic tetramer.

The formation of a cyclic dimer during the polymerization of styrene sulfide has been reported by Noshay and Price (*36*). These authors reported the formation of 2,5-diphenyl-1,4-dithiane in 15% yield when the polymerization was initiated with dimethyl sulfate at room temperature. The following mechanism was proposed:

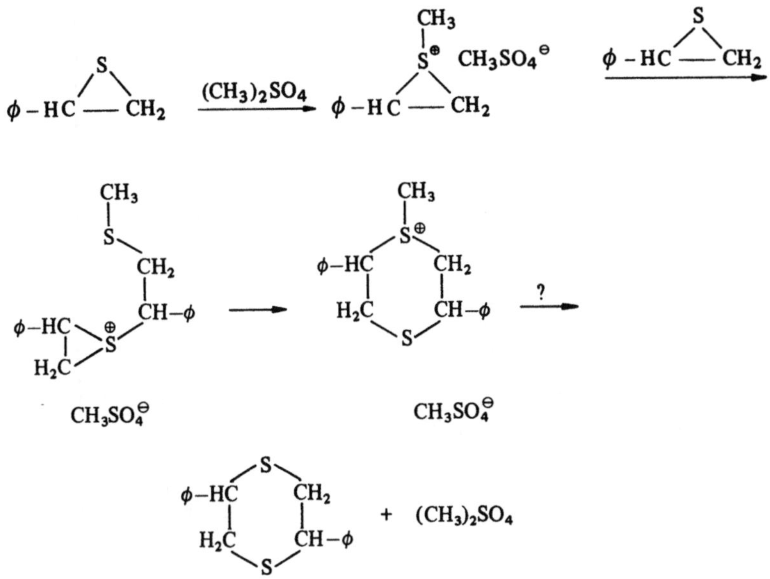

The last reaction, however, (formation of dimethyl sulfate from the methylsulfonium salt) is highly improbable.

The cationic polymerization of styrene sulfide has been reinvestigated by Van Craeynest (*35*). With triethyloxonium tetrafluoroborate as initiator, a rapid and quantitative polymerization was observed, followed by a slow degradation of the polymer to a mixture of *cis* and *trans* 2,5-diphenyl-1,4-dithiane and *cis* and *trans* 2,6-diphenyl-1,4-dithiane. Since the BF_4^- counter ion is not capable of forming a covalent bond, a back-biting reaction via sulfonium ions seems the plausible mechanism for the dimer formation. The polymerization initiated with dimethyl sulfate showed the same characteristics: a fast polymerization is followed by degradation to the same mixture of isomeric diphenyl-1,4-dithianes. However, the *trans*-2,5-diphenyl derivative was the only isomer that crystallized from the solution. It is therefore reasonable to accept that with dimethyl sulfate also, the cyclic dimers of styrene sulfide are formed by a back-biting type of degradation of the polymer and not by the mechanism shown above.

With triethyloxonium tetrafluoroborate, *trans*-2,3-butylene sulfide polymerizes quantitatively within a few minutes but the polymer rapidly degrades to form equimolar amounts of 3,4,6,7-tetramethyl-1,2,5-trithiacycloheptane and *trans*-butene. Figure 3 shows some typical gel permeation chromatograms obtained with this system. The peak at an elution volume of 33 ml is polymer, the positive peak at an elution volume of 52 ml is the trithiacycloheptane derivative, the negative peak at an elution volume of 62 ml is *trans*-butene. *Trans*-2,3-butylene sulfide has a threo configuration (RR or SS). If the polymerization occurs via an S_N2 attack of the sulfide function of a monomer molecule on the α-carbon of a three-membered ring sulfonium salt, one of the two asymmetric carbons will undergo an inversion of configuration leading to a polymer in which the units have erythro configurations (RS or SR, meso form). Considering the fact that in the transition state of elimina-

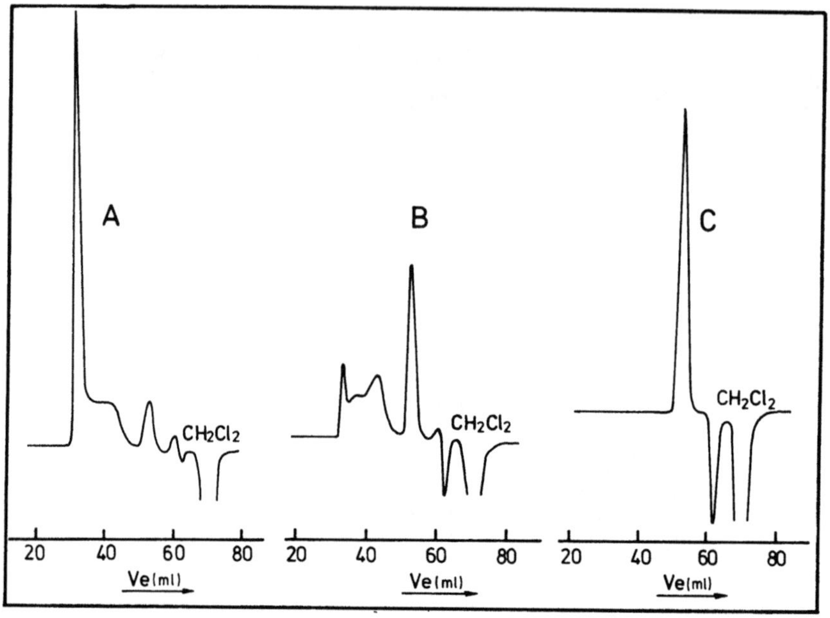

Fig. 3. Gel permeation chromatograms of the "polymerization" of trans-2,3-dimethylthiirane at 20 °C in methylene chloride with triethyloxonium tetrafluoroborate (*35*).
Initial monomer conc.: 1.0 mol · l^{-1}, initiator conc.: 0.01 mol · l^{-1}. A: 20 mins after initiation; B: 80 mins after initiation; C: 1400 mins after initiation

tion reactions the two leaving groups generally must occur in an anti-position to each other, the following mechanism is in agreement with the observation that in the degradation reaction only *trans*-butene and no *cis*-butene is formed:

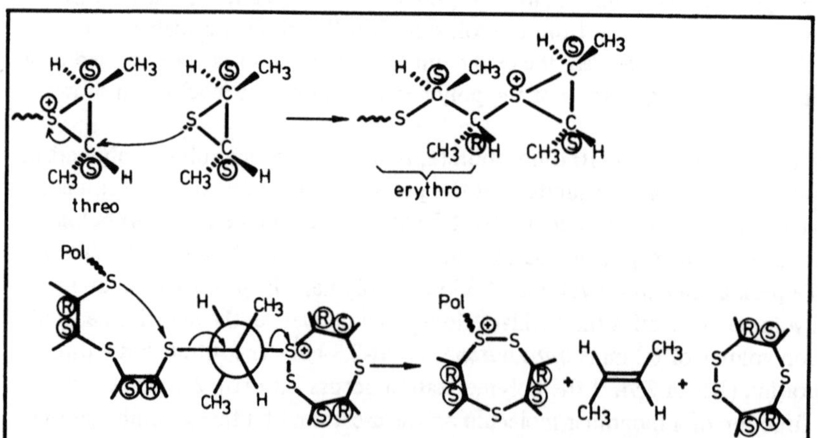

Mechanism of formation of trithiacycloheptane derivative and *trans*-butene by polymerization-degradation of trans(2,3-dimethyl thiirane)

The polymer of *cis*-2,3-butylene sulfide forms a mixture tetramer, the trithiacycloheptane derivative and *cis*-butene. The exclusive formation of *cis*-butene can be explained in a similar way as described previously for the formation of *trans*-butene from *trans*-2,3-butylene sulfide.

Table 3. Cyclic oligomer formation during the cationic polymerization of cyclic sulfides in CH_2Cl_2 as solvent

Monomer	Initiator	Temp. °C	Structure of Oligomer	Ref.
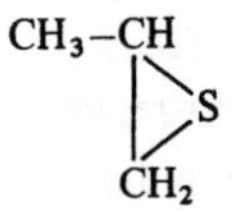	BF_3OEt_2 Et_3OBF_4	0–20	Tetramer, pentamer, dimethyl-1,2,5-trithia-cycloheptanes and propene	(*33, 65*)
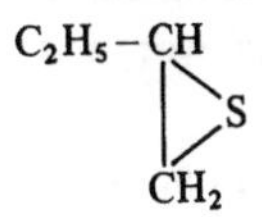	Et_3OBF_4	20	Tetramer	(*35*)
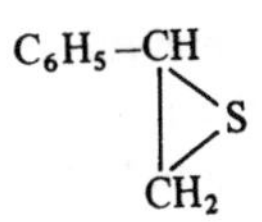	Et_3OBF_4 $(CH_3)_2SO_4$	0–20	Dimer	(*35, 36*)
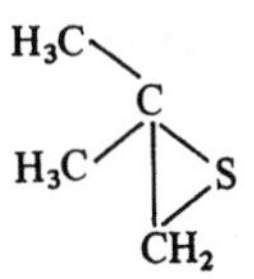	Et_3OBF_4	0–20	Dimer	(*35*)
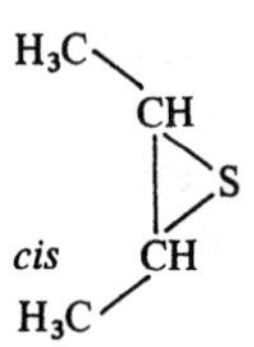	Et_3OBF_4	0–20	*cis*-Butene 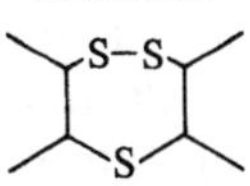 Tetramer	(*35*)
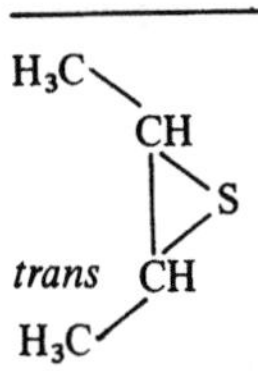	Et_3OBF_4	0–20	*trans*-Butene 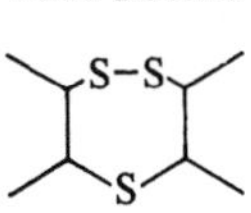	(*35*)
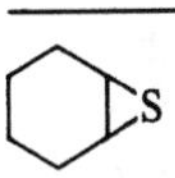	Et_3OBF_4	0–20	Dimer	(*35*)
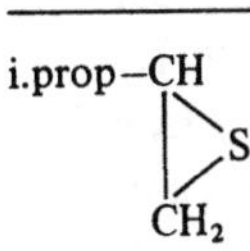	Et_3OBF_4	0–20	Tetramer	(*35*)

3,4,6,7-Tetramethyl-1,2,5-trithiacycloheptane can occur as six geometrical isomers. Two examples are:

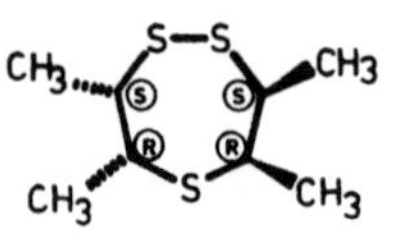

cis-trans-cis

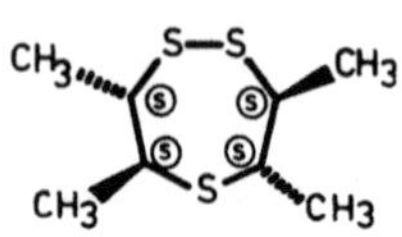

trans-trans-trans

According to the polymerization-degradation mechanism shown above, *trans*-butylene sulfide can lead to only two out of the six geometrical isomers, namely, the *cis-cis-cis*- and the *cis-trans-cis* isomers. *Cis*-butylene sulfide can lead only to the *trans-cis-trans* and *trans-trans-trans* isomers. It was indeed found that the trithiacycloheptane derivatives consisted of a mixture of two isomers and that those derived from the *cis*-monomer were different from those derived from the trans-monomer. It was, however, not possible to attribute an absolute structure to each of the isomers.

The assumption that the disulfide-containing oligomers are formed by a back-biting degradation mechanism of the polymer is confirmed by the observation that high molecular weight poly(*cis*-2,3-butylene sulfide) or poly (*trans*-2,3-butylene sulfide), obtained by ionic coordinative polymerization, are also degraded to the same mixtures of oligomers when treated with a catalytic amount of triethyloxonium tetrafluoroborate.

The formation of an olefinic compound and a 1,2,5-trithiacycloheptane derivative is also observed in the case of tert. butyl thiirane and, to a lesser extent, with cyclohexene sulfide.

With all thiiranes, addition of new monomer to a degraded reaction mixture resulted in an instantaneous start of polymerization of the newly added monomer, followed by degradation. The degradation can be stopped at any moment by adding a small amount of tetrahydrothiophene to the reaction mixture. This sulfide is known to react rapidly with sulfonium salts to form a S-alkyl tetramethylenesulfonium salt (*37*). The resulting reaction mixture is then no longer able to reinitiate polymerization of newly added monomer.

Table 3 gives a survey of the cyclic oligomers formed in the cationic polymerization of thiirane monomers.

5. Aziridines

Dick (*38, 39*) has shown that a number of l-alkylaziridines are converted into the corresponding 1,4-dialkylpiperazines in yields as high as 95%, when treated with hydrohalic acids in acetone or methyl ethyl ketone. Perchloric acid and p-toluenesulfonic acid resulted in the formation of poly(l-alkylaziridines) rather than piperazines. The major products formed in a water solvent were the polymers with 1,4-dialkylpiperazines being formed in maximum yields of 27% in the case of hydroiodic acid. The mechanism proposed by Dick is, that the piperazine are formed via a dimer in which the halogen atom of the initiating acid has formed a covalent bond with a carbon atom. This covalent bond would not be formed with the perchloric acid or with *p*-toluenesulfonic acid, resulting in production of polymer exclusively.

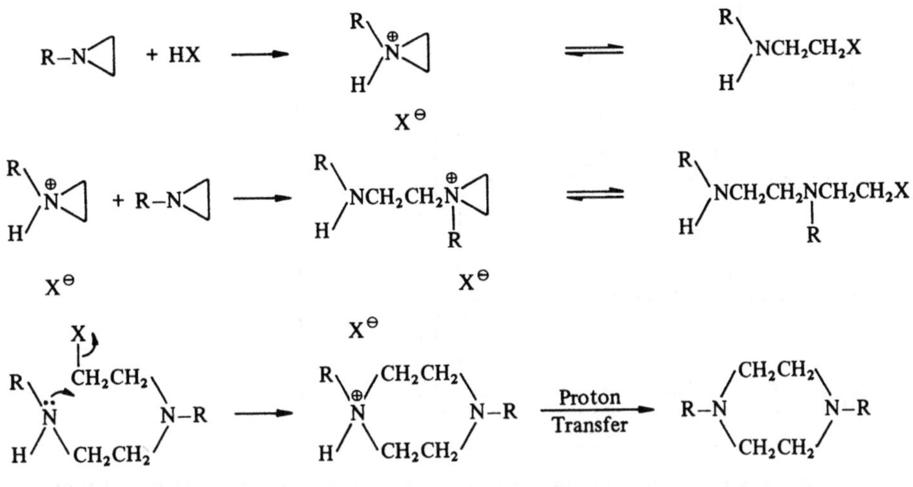

Piperazine formation when X = I, Br, Cl Polymerization when X = TosO, ClO_4

Mechanism of 1,4-dialkylpiperazine formation from l-alkylaziridines (*39*)

When instead of the acids, alkyl halides or methyl *p*-toluenesulfonate were used as the initiator, the dimerization reaction stopped at the N-alkyl piperazinium salt stage (*38*). Here also, dimer formation is favored in acetone while solvents of high polarity and solvating power tend to favor polymer formation.

This effect of solvent polarity on the product distribution is in agreement with the proposed reaction scheme. A solvent such as acetone favors a reaction involving charge destruction, such as the alkyl halide formation from the aziridinium ion with the counter ion. Thus acetone favors piperazine formation. A solvent of high polarity favors charge stabilization or charge transfer such as the polymerization steps. Therefore, polymer or piperazine derivatives can be prepared by the proper choice of solvent.

When 1-phenyl aziridine or 1-(*p*-tolyl)aziridine is treated with sodium iodide in boiling ethanol-water solution the corresponding disubstituted piperazines are formed (*62*). In the absence of the salt only polymerization took place. The piperazines are most probably formed by the same mechanism as proposed by Dick.

Dimer formation from N-alkylaziridines must not necessarily occur via a covalent bond formation with the counter ion. Bruggeman found that a number of l-substituted aziridines form a mixture of polymers and the corresponding piperazine derivatives when treated with triethyloxonium tetrafluoroborate in methylene chloride (*40*). The first reaction product is polymer which degrades more or less slowly to form the cyclic oligomer. As an example Fig. 4 shows two gel permeation chromatograms obtained during the polymerization-degradation process with N-(2-phenylethyl)aziridine. It is clearly seen that the residual monomer concentration (elution volume at 59 ml) remains unchanged and that the cyclic dimer (elution volume 51 ml) is formed by degradation of the polymer (elution volume 35 ml). Since BF_4^- ion is not capable of forming a covalent bond, degradation of polymer involving quaternary ammonium salts is the

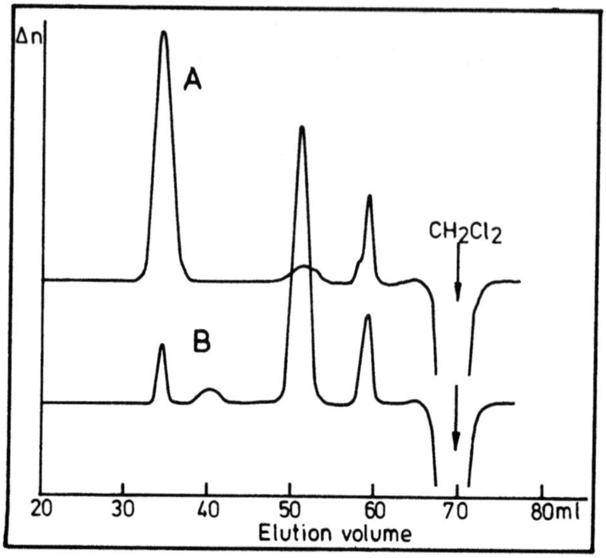

Fig. 4. Gel permeation chromatograms showing the polymerization and degradation of N-(2-phenylethyl)aziridine in methylene chloride with triethyloxonium tetrafluoroborate (*40*). Initial monomer conc.: 1.0 mol · l^{-1}, initiator conc.: 0.0125 mol · l^{-1}. A: 12 hrs after initiation (at 0 °C); B: 120 hrs after initiation (at 20 °C)

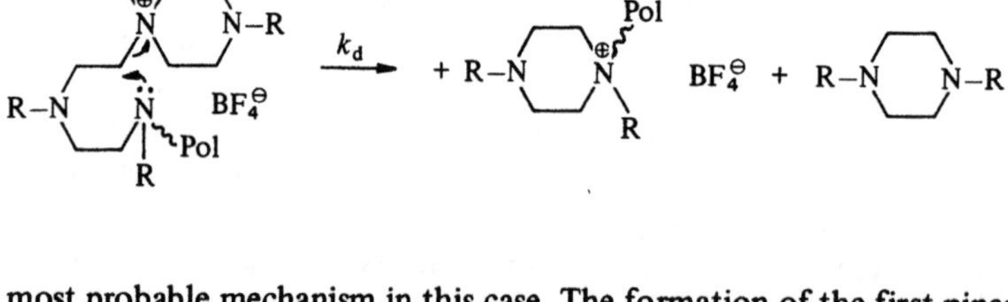

most probable mechanism in this case. The formation of the first piperazinium salt may be regarded as a termination reaction for the propagation.

Treatment of 1-(2-triethylsilylethyl)aziridine with catalytic amounts of aluminum trichloride in boiling heptane gave the corresponding 1,4-disubstituted piperazine in 90% yield (*60*). 1-Methyl aziridine produced a small amount (approximately 10% yield) of dimethylpiperazine only when a large amount of perchloric acid (67 mole %) was added to the monomer and the mixture was allowed to stand for 18 hrs (*61*). In these cases also, the oligomer formation most probably occurs by a backside attack of the penultimate nitrogen, since it is difficult to define a mechanism involving a covalent bond with the counter ion with the used initiator systems.

Table 4. Cyclic oligomer formation during the cationic polymerization of aziridines

Monomer	Initiator	Solvent	Temp. °C	Structure of oligomer	Ref.
CH_2 N–Et CH_2	HI, HBr, HCl	Acetone	25–30	Dimer	(*39*)
CH_2 N–Bu CH_2	HI, HBr, HCl	Acetone	25–30	Dimer	(*39*)
CH_2 N–C_6H_5 CH_2	NaI	H_2O/EtOH	80	Dimer	(*62*)
CH_2 N–C_6H_4–CH_3(p.) CH_2	NaI	H_2O/EtOH	80	Dimer	(*62*)
CH_2 N–CH_2CH=CH_2 CH_2	HI, HBr, HCl	Acetone	25–30	Dimer	(*39*)
CH_2 N–CH_2CH_2–ϕ CH_2	HI, HBr, HCl Et_3OBF_4	Acetone CH_2Cl_2	25–30 20	Dimer Dimer	(*39*) (*40*)
CH_2 N–CH_2CH_2CN CH_2	Et_3OBF_4	CH_2Cl_2	20	Tetramer	(*40*)
CH_2 N–$CH_2\phi$ CH_2	Et_3OBF_4 TosOH	CH_2Cl_2 EtOH	20 75	Dimer Tetramer	(*40*) (*41*)
CH_2 N–$CH_2CH_2Si(C_2H_5)_3$ CH_2	$AlCl_3$	Heptane	98	Dimer	(*60*)
C_2H_5–CH N–$CH_2\phi$ CH_2	BF_3OEt_2	C_6H_6 or EtOH	 25	Tetramer	(*32*)
CH_3–CH N–$CH_2\phi$ CH_2	HI	Acetone		Dimer	(*44*)

Whether an aziridine forms a polymer or an oligomer or a mixture of the two reaction products depends on small changes in monomer structure, on the initiator, and on the reaction conditions. l-Benzylaziridine, for example, is reported to form a cyclic tetramer in nearly quantitative yields with *p*-toluenesulfonic acid in boiling ethanol (*41*). The same monomer leads to a polymer which partially degrades to cyclic dimer when treated with triethyloxonium salts in methylene chloride at 20 °C (*40*). l-Benzyl-2-methylaziridine does not form cyclic oligomers with triethyloxonium tetrafluoroborate (*40*) but l-benzyl-2-(R)ethylaziridine forms a mixture of polymer and of tetramer in a ratio depending on the temperature when treated with borontrifluoride etherate (*42*). l-Ethylaziridine (*39*), l-butylaziridine (*39*), l-allylaziridine (*39*) and l-(2-phenylethyl) aziridine (*39, 40*) form dimers. However, 1-(2-hydroxyethyl)aziridine did not form the corresponding piperazine (*39*), whereas 1-(2-cyanoethal)-aziridine forms a mixture of polymer and tetramer (*40*). 1-Isopropylaziridine, l-tert.butylaziridine and l-phenylaziridine do not form oligomers when treated with triethyloxonium tetrafluoroborate in methylene chloride (*43*).

When chiral N-substituted propyleneimines are treated with hydrogen iodide in acetone at room temperature the corresponding chiral piperazines are obtained (*44*).

Table 4 gives a survey of cyclic oligomers formed in the cationic polymerization of N-substituted aziridines.

C) Discussion

1. General Remarks

Many of the cyclic oligomers reported in the literature have been discovered by chance, for example, because the oligomer formed was a crystalline precipitate or because it was present in such great quantities that it could not be overlooked. Only in the last few years, thanks to the introduction of gel permeation chromatography techniques, has it been possible to analyze more accurately polymer mixtures for the presence of oligomers. Also, except in a few cases, it is not known whether the amount of oligomer reported in a paper is invariably the same or if it is a fortuitous amount typical for a particular experiment. A striking example is that of the 2,3-butene episulfides. Under the influence of triethyloxonium tetrafluoroborate reaction at 0 °C for a short reaction time, yields polymer in quantitative yields. However, after a few hours the polymer is already partially degraded and if the experiment were carried out at room temperature using "normal" polymerization times, only cyclic oligomers would be obtained (*35*). So the question remains open as to what is the amount of oligomer formed with monomers such as styrene oxide, propylene oxide etc. under reaction conditions other than those described in the literature. It is also very probable that other monomers which have not been reported to form oligomers, in fact do form oligomers but that these have not been observed because they coprecipitated with the polymer or were eliminated during the purification of the polymer.

2. Reaction Mechanisms

A number of authors propose a mechanism for oligomer formation. The mechanisms can be divided into two substantially different groups: (a) oligomer formation is a side reaction of the polymerization, which means that oligomer is formed directly from monomer; (b) oligomers are formed by degradation of polymer.

a) Oligomer is Formed Directly from Monomer

Here we can again make a distinction between: (i) oligomerization occurs by the same mechanism as polymerization and (ii) oligomerization follows a different mechanism. The first kind has been proposed by Rose (*19*) for the tetramer formation of oxetane: up to the fourth monomer added there is no difference between polymerization and tetramer formation. For polymerization the next step is addition of a fifth monomer molecule. If the next step is reaction between the end-standing hydroxy group with the growing chain, a cyclic secondary oxonium salt is formed which leads to tetramer by a proton transfer to another ether function (see p. 109).

A rather clear situation exists in the mechanism for piperazine formation from N-alkylaziridines proposed by Dick (*38, 39*) (see p. 119). The nucleophilic character of the counter ion determines if dimer or polymer is formed after two reaction steps starting from monomer.

Oligomer formation by a mechanism different from the polymerization mechanism has been proposed by Kern (*7*) for the tetramer formation from a number of epoxides and by Pasika (*10*) for styrene oxide dimer formation, although in the last case the initiation reactions are probably the same. However, in both cases degradation of polymer would be a possible alternative mechanism.

A system where good evidence exists that oligomers are formed directly from monomer is that of the cationic polymerization of hexamethylcyclotrisiloxane (45). Apart from polymer of molecular weight 10^5, a mixture of all cyclic siloxanes with ringsize from 6 (monomer) up to 48 have been detected using gaschromatography. The important observation was that oligomers with ringsizes 6, 12, 18, 24 . . . etc. were 10 to 100 times more abundant than the others. This strongly indicates that these oligomers are formed directly from monomer since there are no reasons for supposing that the formation of these rings by a back-biting mechanism from polymer would be especially favored.

b) Oligomer is Formed by Degradation of Polymer

If we consider ring-opening polymerization as equilibrium reactions between monomer and polymer, the equilibrium monomer concentration $[M_e]$ will be given by

$$\ln[M_e] = \frac{\Delta G^\circ}{RT} = \frac{\Delta H^\circ}{RT} - \frac{\Delta S^\circ}{R}$$

where ΔG° is the free energy change for the transformation of monomer to polymer (*46*). In an article written in 1955, Dainton, Devlin and Small (*47*) have calcu-

Table 5. Estimated theoretical equilibrium monomer concentrations at 300 °K

Monomer	Ringstrain[a] (kcal · mol^{-1})	ΔG_p° [b] at 300 K	$[M]_e$ at 300 K[c] (mol · l^{-1})
	27	−21	6.10^{-16}
	20	−14	7.10^{-11}
	25.5	−16.5	10^{-12}
	20	−14	7.10^{-11}
	5.6	+0.4	1.9
	2.0	+4	8.10^{2}
	−0.3	+6.3	$3.6 \cdot 10^{4}$

[a] Pell, A. S. and Pilcher, G.: Trans. Faraday Soc., 61, 71 (1965).
[b] Assuming $-\Delta H_p^\circ$ = ringstrain and $\Delta S_p^\circ = -20$ cal K^{-1}mol^{-1}.
[c] $[M]_e = \exp(\Delta G_p^\circ/RT)$.

lated that for the (hypothetical) polymerization of cycloparaffins and their methyl and 1,1-dimethyl derivatives, the free energy of polymerization is negative at least up to cyclo-octane with the exception of cyclohexanes and substituted cyclopentanes. In a subsequent article Small (*48*) estimated that the conclusions arrived at for the cycloparaffins will hold, at least qualitatively for heterocyclic compounds containing O or NH. Some deviations are expected for the sulfur compounds due to the larger bond distance of C–S compared to C–C, C–O and C–N. Since then, the values for the enthalpy (ΔH°), entropy (ΔS°) and free energy of polymerization for a number of ring-opening polymerizations have been determined (*49*). For three- and four-membered rings, the enthalpy of polymerization is generally around −20 kcal mol^{-1}, whereas for most unsubstituted five-membered rings it is around −5 kcal · mol^{-1}. The entropy of polymerization is always negative and the observed values generally lay between −10 and −25 cal K^{-1} mol^{-1}.

Assuming that the enthalpy of polymerization is the negative of the ringstrain and taking for ΔS_p° a value of −20 cal K^{-1} mol^{-1}, the free energy of polymerization can be estimated and therefrom the equilibrium monomer concentration. Some examples are given in Table 5. From this table it follows that for three- and four-membered ring monomers, ΔG° at room temperature has highly negative values, so that the equilibrium monomer concentrations in those polymerizations are extremely low. For the five-membered ring tetrahydrofurane, the factor $T\Delta S°$ becomes of the same order of magnitude as $\Delta H°$ at temperatures around room temperature, so that measurable equilibrium monomer concentrations are found. These equilibrium concentrations increase with temperature and at a certain temperature they will reach a value equal to the molar concentration of the pure monomer. Above this temperature polymerization is thermodynamically impossible and on the contrary, the polymer will depolymerize, provided that a suitable mechanism exists to allow this depolymerization. For six-membered and some five-membered rings, $\Delta H°$ for polymerization is close to zero, or even slightly positive, with the effect that at normal temperatures $\Delta G°$ for polymerization will be positive. Inversely, $\Delta G°$ for depolymerization will have the same value but be negative. In other words, these monomers will be formed from the polymer, if a suitable mechanism is provided. This is well illustrated by the case of tetrahydrothiophene. This monomer cannot be polymerized at ambient temperatures because ΔG_p°, calculated from a ringstrain of 2.0 kcal (*50*) and assuming ΔS_p° equal to -20 cal K^{-1} mol^{-1} has a value of +4.0 kcal at 300 °K which would correspond to an equilibrium monomer concentration of more than 10^2 mol · 1^{-1}, a value that can never be obtained. Inversely, poly(tetramethylene sulfide), prepared by polycondensation, should degrade to form tetrahydrothiophene when the reactive species, necessary to allow this degradation, are introduced in the polymer. It was indeed shown that addition of a catalytic amount of triethyloxonium tetrafluoroborate to a solution of poly(tetramethylenesulfide) resulted in degradation of the polymer with formation of tetrahydrothiophene (*51*).

All polymers derived from three-membered heterocycles have structures which allow the formation of six-membered rings. Therefore, it is thermodynamically expected that these polymers will degrade, as long as active species which allow these degradations to occur, are present in the polymer chains. This explains why polyoxiranes degrade to 1,4-dioxanes, polythiiranes to 1,4-dithianes and poly(N-alkylaziridines) to dialkylpiperazines.

Good evidence exists that cationic polymerization of heterocyclic monomers occurs with cyclic onium salts as the active species. Therefore it is reasonable to believe that also the degradation reactions occur via such ions, so that the most plausible reaction path is the classical back-biting mechanism:

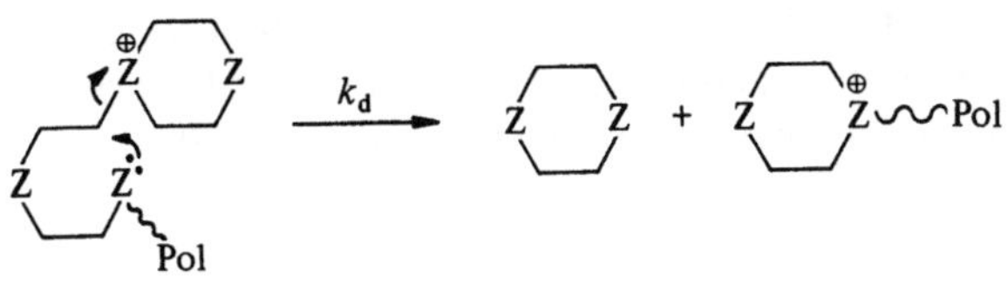

In fact the problem with these polymerizations is not so much: "why does degradation occur? " but rather "why do some polymers not degrade? " and also "why do some polymers degrade to other rings than six-membered rings? "

An obvious explanation for the fact that degradation does not occur under polymerization conditions is that the free energy of activation for the degradation is too high so that the rate of degradation is too low to observe the phenomenon. This free energy of activation can be influenced by the nature of the solvent and the counter ion which might explain the different behavior observed for some monomers when polymerized under different conditions. The presence of substituents on the monomer will certainly play an important role in this connection. However, at the present time the number of experimental data are insufficient to allow general conclusions about the relationship between the nature and the number of substituents and the tendency to degradation.

Degradation of polymer may be absent or occur only to a small extent because of the occurrence of a termination reaction for the degradation. This termination reaction may or may not be the same as the termination reaction for the propagation. The transformation of a three-membered cyclic onium salt to a six-membered cyclic onium salt for example, would be a termination reaction for the propagation, but can be the first step of the degradation process. Such a situation is believed to exist in the polymerization of some N-substituted aziridines (*40*): polymerization stops at limited conversions but degradation of the polymer to the corresponding piperazine derivative continues after the polymerization has stopped and this in the presence of unreacted monomer (see p. 120). Termination of both polymerization and degradation reactions has been observed for the polymerization of propylene sulfide with triethyloxonium hexachloroantimonate (*52*). The termination in this case is believed to be a reaction between the active sulfonium ion and the counter ion forming a chain end which is non-active for further propagation or for degradation.

Sometimes oligomers are formed only during the polymerization and when polymerization stops, because of complete monomer consumption, also the formation of the oligomer ceases. This behavior has been observed with the oxetanes (see p. 109). At least three mechanisms can explain this kind of situation.

(i) Polymerization and oligomer formation occur via two independent reaction paths and both reactions stop when monomer is exhausted.

(ii) Propagation and degradation of the polymer to form oligomer occur simultaneously at approximately the same rate. Normally this would lead to oligomer as the only reaction product. However, if a termination reaction prevents the complete degradation of the polymer the end product will be a mixture of polymer and oligomer.

(iii) Oligomer is formed by degradation of polymer but monomer is involved in the mechanism of oligomer formation, in other words, oligomer is formed from the polymer, but only in the presence of monomer. This kind of reaction has been proposed for the tetramer formation in the polymerization of oxetanes (*53*):

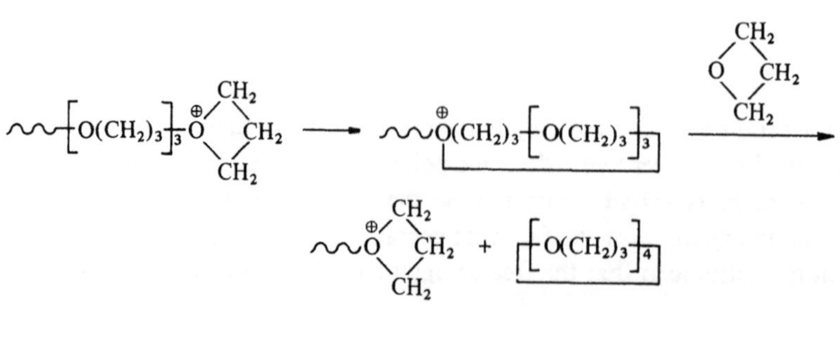

According to this mechanism, after quantitative polymerization, oligomer formation would not be possible any more because of the absence of monomer. At the present time it is not possible to decide by which mechanism the tetramers of oxetane and 3,3-dimethyl oxetane are formed.

Many polymers derived from three-membered heterocycles do not degrade to six-membered rings but to twelve-membered or to larger rings. If cyclic oligomers are formed by a back-biting mechanism it is difficult to explain why these 12-membered rings are formed with such a high preference. Considering the results of the degradation studies with mixtures of deuterated and non-deuterated poly(propylene sulfide), as described in Section 4, p. 114, the most plausible mechanism for the formation of the tetramers is, that there is a continuous exchange between sulfonium salts and sulfide functions which ultimately leads to the thermodynamically most stable compound. It is possible that an analogous mechanism operates in the case of propylene oxide. Entelis and co-workers (*17*) have reported that the cyclic tetramer of propylene oxide is thermodynamically more stable than the corresponding dimer or polymer which is an indication that the above mentioned mechanism is valuable. This would lead to the conclusion that if dimer formation by a back-biting mechanism is slow, because of a high activation energy, tetramer is formed by a thermodynamically controlled reaction. In a back-biting mechanism the transition state for the formation of a six-membered ring can tentatively be represented as follows:

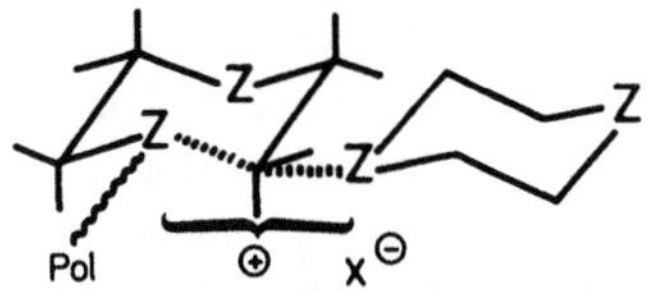

If one or more of the atoms in the ring to be formed carry substituents instead of hydrogen atoms, at least some of these substituents will occur in a gauche butane conformation either with the polymer chain or with the leaving group or/and because they occupy an axial position on the ring. Since each gauche butane interaction increases the enthalpy of activation of ring formation with 0,8 kcal · mol^{-1}, the occurence or absence of such interactions must have an influence on the rate of oligomer formation. It is expected that this effect is less pronounced when a twelve-membered ring is formed because in that case the ring can adopt such conformations where these gauche butane interactions do not occur.

The importance of the configuration of the recurring units of the polymer chain on the degradation path is well demonstrated by comparison of the behaviour of poly(*cis*-2,3-butylene sulfide) and that of poly(*trans*-2,3-butylene sulfide). Both polymers degrade rapidly at room temperature but the degradation products are different. Yet, the only difference in the polymer structure is, that the recurring units of the polymer derived from the cis-isomer have threo configurations, whereas those of the polymer derived from the trans-isomer have erythro configurations. A similar observation has been made with propylene sulfide: Isotactic poly(propylene sulfide) degrades quantitatively to oligomers, predominantly tetramer, when treated with a catalytic amount of oxonium salt (*28*). Poly(propylene sulfide) obtained by cationic polymerization on the other hand, does not degrade completely (*27*). This might be caused by the presence of a number of head-to-head structures in this polymer (*54*).

D) Conclusion

The formation of cyclic oligomers during or after the cationic polymerization of heterocyclic monomers is a frequently occuring phenomenon. Mostly, the oligomers are formed by degradation of polymer and not directly from monomer. The degradations occur via onium ions which are present in the reaction mixtures because the polymerizations also occur via onium salts. Degradations stop when the active onium salts are destroyed. This can be due to a termination reaction or can be caused by adding a stop reagent.

With a limited number of monomers, oligomers are formed only during the polymerization and not after. This would mean polymer and oligomer are formed by two independent reaction paths or that a termination reaction exists for the degradation, or that the monomer is involved in the degradation reaction which produces the oligomer. In other cases the degradation reaction continues after the polymerization has stopped and stops before all polymer is degraded for reasons which are unknown at present. From the few studies reported up to now, it seems as if there are two kinds of degradation reactions. The first kind is a simple back-biting reaction comparable with the depolymerization reaction for polytetrahydrofurane. The second kind would be the result of a great number of transalkylation reactions leading to the thermodynamically most stable compounds. For entropy reasons and possibly also for enthalpy reasons, these compounds appear to be cyclic oligomers.

The ringsize of the cyclic oligomers depends on a number of factors. Steric hindrance by substituents, temperature, nature of the counter ion and solvent seem to play an important role in this connection. However, the number of systematic studies in this field is too small to allow general rules.

The knowledge of the mechanism by which cyclic oligomers are formed during the cationic polymerization of heterocycles is important in order to suppress this reaction or to favour it. If one wishes to produce a polymer the oligomers are undesirable byproducts. On the other hand, it may be interesting from a synthetic point of view to find the experimental conditions which lead to high yields of these oligomers. Finally, in those cases where degradation is a clear-cut back-biting reaction, the study of the stereochemistry of the oligomers formed from polymers may help to elucidate the stereochemistry of the polymer and hence give information about the mechanism of polymerization reactions. This applies not only to cationic polymerizations, since a great number of polymers formed by other mechanisms than cationic, can be degraded to oligomers when treated with cationic catalysts.

E) References

1. Pedersen, C. J., Frensdorf, H. K.: Angew. Chem., Internat. Edn. **11**, 16 (1972).
2. Worsfold, D. J., Eastham, A. M.: J. Am. Chem. Soc. **79**, 897 (1957).
3. Worsfold, D. J., Eastham, A. M.: J. Am. Chem. Soc. **79**, 900 (1957).
4. Latremouille, G. A., Merrall, G. T., Eastham, A. M.: J. Am. Chem. Soc. **82**, 120 (1960).
5. Dale, J., Borgen, G., Daasvatn, K.: Ger. Offen. 2,401, 126 (July 18th, 1974).
6. Katnik, R. J., Schaefer, J.: J. Org. Chem. **33**, 384 (1968).
7. Kern, R. J.: J. Org. Chem. **33**, 388 (1968).
8. Colclough, R. O., Wilkinson, K.: J. Polym. Sci., Part C, **4**, 311 (1966).
9. Colclough, R. O., Oze, G., Higginson, W. C. E., Jackson, J. B., Litt, M.: J. Polym. Sci. **34**, 171 (1959).
10. Pasika, W. M.: J. Polym. Sci. **A-3**, 4287 (1965).
11. Kondo, S., Blanchard, L. P.: Polymer Letters **7**, 621 (1969).
12. Eastham, A. M. in: The chemistry of cationic polymerization. (P. H. Plesch, ed.), Chap. 10. New York: Pergamon Press 1963.
13. Vandenberg, E. J.: J. Polym. Sci., Part A-1, **10**, 329 (1972).
14. Weissermel, K., Nölken. E.: Makromol. Chem. **68**, 140 (1963).
15. Entelis, S. G., Korovina, G. V., Kuzayev, A. I.: Vysokomol. Soed. **A-13**, 1438 (1971).
16. Estrin, Y. I., Entelis, S. G.: Vysokomol. Soed. **A-13**, 1654 (1971).
17. Entelis, S. G., Korovina, G. V.: Makromol. Chem. **175**, 1253 (1974).
18. Tsuruta, T.: private communication.
19. Rose, J. B.: J. Chem. Soc. **1956**, 542 and 546.
20. Arimatsu, Y.: J. Polym. Sci. **A-1**, 728 (1966).
21. Dreyfuss, P., Dreyfuss, M. P.: Polymer Journ. **8**, 81 (1976).
22. Bucquoye, M., Goethals, E. J.: unpublished results (1975).
23. Andrews, J. M., Semlyen, J. A.: Polymer **12**, 642 (1971).
24. Andrews, J. M., Semlyen, J. A.: Polymer **13**, 142 (1972).
25. Miki, T., Higashimura, T., Okamura, S.: Polymer Letters, **5**, 65 (1967).
26. S. Okamura, S., Higashimura, T., Miki, T.: Progress in polymer science, Japan, Vol. *3*, 97 (1972).
27. Miki, T., Higashimura, T., Okamura, S.: J. Polym. Sci. Part A-1, **5**, 2977 (1967).
28. Miki, T., Higashimura, T., Okamura, S.: Polym. Sci. Part A-1, **5**, 95 (1967).
29. Higashimura, T., Tanaka, A., Miki, T., Okamura, S.: J. Polym. Sci. Part A-1, **5**, 1937 (1967).
30. Plesch, P. H., Westermann, P. H.: Polymer **10**, 105 (1969).
31. Okada, M., Kozawa, S., Yamashita, Y.: Makromol. Chem., **127** 271 (1969).
32. Van Ooteghem, D., Goethals, E. J.: Makromolek. Chem., **175** 1513 (1974).
33. Lambert, J. L., Van Ooteghem, D., Goethals, E. J.: J. Polym. Sci. Part A-1, **9**, 3055 (1971).
34. Goethals, E. J., Van Ooteghem, D., Van Craeynest, W.: Preprint nr. C-12 of the International Symposium on Cationic Polymerization, Rouen, September 1973.
35. Van Craeynest, W., Goethals, E. J.: Europ. Polym. J., **12**, 859 (1976).
36. Noshay, A., Price, C. C.: J. Polym. Sci. **54**, 533 (1961).
37. Van Ooteghem, D., Deveux, R., Goethals, E. J.: Int. J. Sulfur Chem. **8**, 81 (1973).
38. Dick, C. R.: J. Org. Chem. **32**, 72 (1967).
39. Dick, C. R.: J. Org. Chem. **35**, 3950 (1970).
40. Goethals, E. J., Schacht, E. H., Bruggeman, P.: Preprint nr. C-30 of the International Symposium on Cationic Polymerization, Rouen, September 1973.
41. Hansen, G. R., Burg, T. E.: J. Heterocycl. Chem. **5**, 305 (1968).
42. Tsuboyama, S. Tsuboyama, K., Higashi, I., Yanagita, M.: Tetrah. Letters, **16**, 1367 (1970).
43. Bossaer, P., Goethals, E. J.: unpublished results.
44. Yanagite, M., Tsuboyama, S., Tsuboyama. K.: Jap. Pat. Kokai 74 109386, C. A. **83** 10142a (1975).
45. Chojnowski, J., Mazurek, M., Scibiorec, M., Wilczek, L.: Makromol. Chem. **175**, 3299 (1974).
46. Dainton, F. S., Ivin, K. J.: Quaterly Reviews **12**, 61 (1958).
47. Dainton, F. S., Devlin, T. R. E., Small, P. A.: Trans. Faraday Soc. **51**, 1710 (1955).

48. Small, P. A.: Trans. Faraday Soc. **51**, 1717 (1955).
49. Allen, P. E. M., Patrick, C. R.: Kinetics, mechanisms of polymerization reactions, p. 211. New York: John Wiley & Sons 1974.
50. Pell, A. S., Pilcher, G.: Trans. Faraday Soc., **61**, 71 (1965).
51. Van Ooteghem, D., Deveux, R., Goethals, E. J.: J. Polym. Sci., in press.
52. Goethals, E. J.: Makromol. Chem. **175**, 1309 (1974).
53. Saegusa, T., Hashimoto, Y., Matsumoto, S.: Macromolecules **4**, 1 (1971).
54. Boileau, S., Cheradame, H., Spassky, N., Ivin, K., Lillie, E.: C. R. Acad. Sci. (Paris) **C275**, 535 (1972).
55. Boehlke, K., Jaacks, V.: Makromol. Chem. **145**, 219 (1971).
56. Jaacks, V.: Makromol. Chem. **101**, 33 (1967).
57. Weissermel, K., Fischer, E., Gutweiler, K., Herman, D. D.: Kunststoffe **54**, 410 (1964).
58. Jaacks, V.: Adv. Chem. Ser. **91**, 371 (1969).
59. Kops, J., Spanggaard, H.: Makromol. Chem. **176**, 299 (1975).
60. Nametkin, N. S., Grushevenko, I. A., Perchenko, V. N.: Dokl. Akad. Nauk. SSSR, Chem. Sect., **162**, 347 (1965).
61. Jones, G. D., MacWilliams, D. C., Braxtor, N.: J. Org. Chem. **30**, 1994 (1965).
62. Heine, H. W., Kenyon, W. G., Johnson, E. M.: J. Am. Chem. Soc. **83**, 2570 (1961).
63. Kern, W., Deibig, H., Giefer, A., Jaacks, V.: Pure Appl. Chem. **12**, 371 (1966).
64. Kobayashi, S., Morikana, K., Saegusa, T.: Macromolecules **8**, 952 (1975).
65. Simonds, R., Goethals, E. J.: to be published.
66. Dale, J., Borgen, G., Daasvatn, K.: Acta Chem. Scand. **B28**, 378 (1974).
67. The formation of isomeric dimethyl-1,2,5-trithiacycloheptanes and propene has also been observed (*65*).

Received February 6, 1976

Author Index Volumes 1–23

B. Vollmert

POLYMER CHEMISTRY

Translated from the German by E. H. Immergut, New York
With 630 figures. XVII, 652 pages. 1973

This book gives a comprehensive coverage of the synthesis of polymers and their reactions, structure, and properties. The treatment of the reactions used in the preparation of macromolecules and in their transformation into crosslinked materials is particularly detailed and complete. The book also gives an up-to-date presentation of other important topics, such as enzymatic and protein synthesis, solution properties of macromolecules, polymer crystallization, and properties of polymers in the solid state.
The content and presentation of Professor Vollmert's book is more encompassing than most existing treatises, and its numerous figures and tables convey a wealth of data, never, however, at the expense of intellectual clarity or educational value.
The presentation is mainly on a fundamental and general level and yet the reader – student or professional – is gradually and almost casually introduced to all important natural and synthetic polymers. Complicated phenomena are explained with the aid of the simplest available examples and models in order to ensure complete understanding. However, the reader is also encouraged to think for himself and even to criticize the author's point of view.
All of the chapters have been revised and enlarged from the German edition, and many of the sections are entirely new.

Contents

Introduction. – Structural Principles. – Synthesis and Reactions of Macromolecular Compounds. – The Properties of the Individual Macromolecule. – States of Macromolecular Aggregation.

REACTIVITY AND STRUCTURE

Concepts in Organic Chemistry

Editors: K. Hafner, J.-M. Lehn, C. W. Rees, P. v. R. Schleyer, B. M. Trost, R. Zahradník

Volume 1
J. Tsuji: **Organic Synthesis by Means of Transition Metal Complexes A Systematic Approach**

Volume 2
K. Fukui: **Theory of Orientation and Stereoselection**

Volume 3
H. Kwart, K. King: **D-Orbital Involvement in the Organo-Chemistry of Silicon, Phosphorus and Sulfur**

LECTURE NOTES IN CHEMISTRY

Editors: G. Berthier, M. J. S. Dewar, H. Fischer, K. Fukui, H. Hartmann, H. H. Jaffé, J. Jortner, W. Kutzelnigg, K. Ruedenberg, E. Scrocco, W. Zeil

Volume 1
G. H. Wagnière: **Introduction to Elementary Molecular Orbital Theory and to Semiempirical Methods**

Volume 2
E. Clementi: **Determination of Liquid Water Structure, Coordination Numbers for Ions and Solvation for Biological Molecules**

**Springer-Verlag
Berlin Heidelberg New York**

GPSR Compliance

The European Union's (EU) General Product Safety Regulation (GPSR) is a set of rules that requires consumer products to be safe and our obligations to ensure this.

If you have any concerns about our products, you can contact us on ProductSafety@springernature.com

In case Publisher is established outside the EU, the EU authorized representative is:

Springer Nature Customer Service Center GmbH
Europaplatz 3
69115 Heidelberg, Germany

Batch number: 10620881

Printed by Printforce, the Netherlands